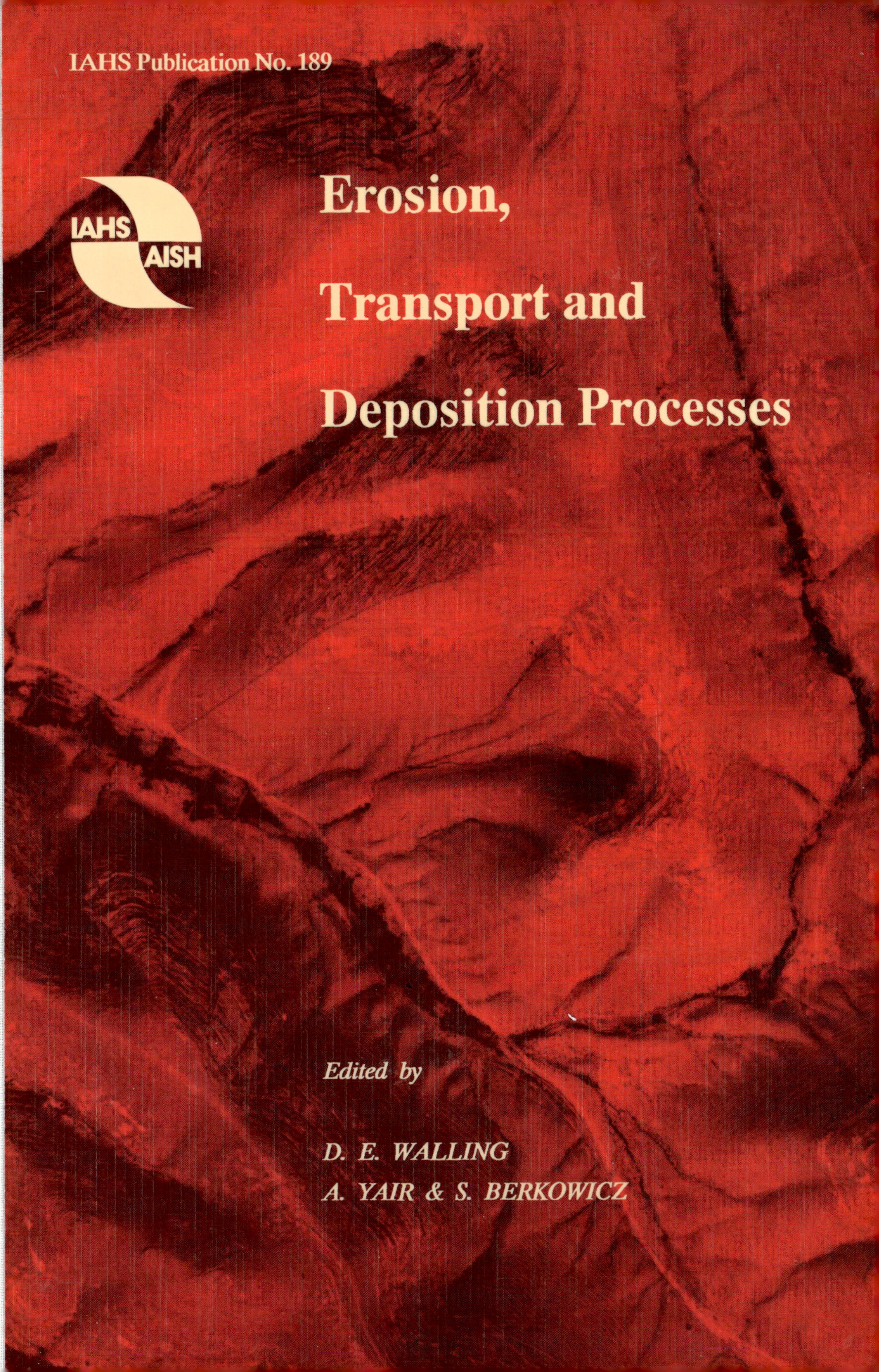

IAHS Publication No. 189

IAHS AISH

Erosion, Transport and Deposition Processes

Edited by

D. E. WALLING
A. YAIR & S. BERKOWICZ

INTERNATIONAL ASSOCIATION OF HYDROLOGICAL SCIENCES

IAHS adheres to the International Union of Geodesy and Geophysics

Officers of IAHS elected for the period 1987-1991

Information

Individual membership of the Association is ***free*** and open to anyone who endeavours to participate in IAHS activities, such as IAHS symposia, workshops and working groups. Further information may be obtained from the Secretary General:

Mr H. J. Colenbrander, Secretary General IAHS, CHO-TNO, PO Box 297, 2501 BD, The Hague, The Netherlands

Information about the activities of the Commissions and the Committee instituted by the Association can be obtained from their Presidents:

Prof. H. J. LIEBSCHER, *President of the IAHS International Commission on Surface Water,* Federal Institute of Hydrology, PO Box 309, D-5400 Koblenz, FR Germany.

Prof. S. M. GORELICK, *President of the IAHS International Commission on Groundwater,* Department of Applied Earth Sciences, Stanford University, Stanford, California 94305, USA.

Prof. D. E. WALLING, *President of the IAHS International Commission on Continental Erosion,* Department of Geography, University of Exeter, Amory Building, Rennes Drive, Exeter EX4 4RJ, UK.

Prof. V. M. KOTLYAKOV, *President of the IAHS International Commission on Snow and Ice,* Institute of Geography, USSR Academy of Sciences, Staromonetny St. 29, Moscow 109017, USSR.

Dr R. A. GRAS, *President of the IAHS International Commission on Water Quality,* EDF, Division des Etudes et Recherches, Service AEE, 6 Quai Watier, 78400 Chatou, France.

Prof. U. SHAMIR, *President of the IAHS Inte tional Commission on Water Resou Systems,* Faculty of Civil Engineer TECHNION, 32000 Haifa, Israel.

Mr J. W. TREVETT, *President of the I International Committee on Remote Sensing Data Transmission,* ISMARSC, 14 Chu Grove, Little Chalfont, Buckinghamshire I 6SH, UK.

Hydrological Sciences Journal

The Association has produced a scient journal since 1956. From 1956 to 19 it was called *Bulletin of the Internatio Association of Scientific Hydrology*, fr 1972 to 1981 the title was changed *Hydrological Sciences Bulletin*, and present title, ***Hydrological Scien Journal***, came into effect from 19 From 1988 the frequency was increas from quarterly to bimonthly. As well scientific papers on all aspects of hydr logy the Journal contains announceme on and news of worldwide hydrologi activities organized or sponsored IAHS, book reviews, and a diary forthcoming events.

Subscriptions. The annual su scription rate for 1990 is £75 ($124 North America). Please send 19 subscriptions to: *Blackwell Scient Publications Ltd, Osney Mead, Oxf OX2 0EL, UK.* The 1978–1989 issu should also be ordered from Blackw Back issues of most volumes prior 1978 are still available and may ordered from any of the addresses proceedings given on the back cov Individual Members of IAHS m obtain a half-price subscription. F further information please cont Sarah Cage (address below).

Contributions. The Editor we comes original papers, scientific not and letters. Please send mater submitted for publication to:

Ms Sarah Cage, IAHS Press, Institute of Hydrol Wallingford, Oxfordshire OX10 8BB, UK

Original papers will be screened by t referees, usually chosen from the Boa of Associate Editors. *Instructions Authors* are available from Sarah Cag

(continued on inside back cover)

EROSION, TRANSPORT AND DEPOSITION PROCESSES

TITLES RECENTLY PUBLISHED BY IAHS

Proceedings of the symposia held during the Second IAHS Assembly, Budapest, July 1986:

Modelling Snowmelt-Induced Processes
Publ.no.155 (1986), price $40

Conjunctive Water Use
Publ.no.156 (1986), price $48

Monitoring to Detect Changes in Water Quality Series
Publ.no.157 (1986), price $40

Integrated Design of Hydrological Networks
Publ.no.158 (1986), price $40

Drainage Basin Sediment Delivery. Proceedings of the Albuquerque Symposium, August 1986
Publ.no.159 (1986), price $45

Hydrologic Applications of Space Technology. Proceedings of the Cocoa Beach Workshop, August 1985
Publ.no.160 (1986), price $45

Karst Water Resources. Proceedings of the Ankara Symposium, July 1985
Publ.no.161 (1986), price $45

Avalanche Formation, Movement and Effects. Proceedings of the Davos Symposium, September 1986
Publ.no.162 (1987), price $50

Developments in the Analysis of Groundwater Flow Systems. Report prepared by a Working Group of the IAHS International Commission on Groundwater
Publ.no.163 (1986), price $35

Water for the Future: Hydrology in Perspective
Proceedings of the Rome Symposium, April 1987
Publ.no.164 (1987), price $50

Erosion and Sedimentation in the Pacific Rim. Proceedings of the Corvallis Symposium, August 1987
Publ.no.165 (1987), price $55

Proceedings of the symposia held during the IUGG Assembly, Vancouver, August 1987:

Large Scale Effects of Seasonal Snow Cover
Publ.no.166 (1987), price $42

Forest Hydrology and Watershed Management
Publ.no.167 (1987), price $55

The Influence of Climate Change and Climatic Variability on the Hydrologic Regime and Water Resources
Publ.no.168 (1987), price $55

Irrigation and Water Allocation
Publ.no.169 (1987), price $32

The Physical Basis of Ice Sheet Modelling
Publ.no.170 (1987), price $40

Hydrology 2000. Report of the IAHS Hydrology 2000 Working Group
Publ.no.171 (1987), price $22

Side Effects of Water Resources Management. Report prepared by an IHP-III Working Group
Publ.no.172 (1988), price $40

Groundwater Monitoring and Management. Proceedings of the Dresden Symposium, March 1987
Publ.no.173 (1990), price $55

Sediment Budgets. Proceedings of the Porto Alegre Symposium, December 1988
Publ.no.174 (1988), price $60

Consequences of Spatial Variability in Aquifer Properties and Data Limitations for Groundwater Modelling Practice. Report prepared by a Working Group of the International Commission on Groundwater
Publ.no.175 (1988), price $45

Karst Hydrogeology and Karst Environment Protection. Proceedings of the IAH/IAHS Guilin Symposium, October 1988
Publ.no.176 (1988), price $55

Estimation of Areal Evapotranspiration. Proceedings of a workshop held during the IUGG Assembly, Vancouver, August 1987
Publ.no.177 (1989), price $45

Remote Data Transmission. Proceedings of a workshop held during the IUGG Assembly, Vancouver, August 1987
Publ.no.178 (1989), price $30

Proceedings of symposia held during the Third IAHS Scientific Assembly, Baltimore, Maryland, May 1989:

Atmospheric Deposition
Publ.no.179 (1989), price $45

Systems Analysis for Water Resources Management: Closing the Gap Between Theory and Practice
Publ.no.180 (1989), price $45

Surface Water Modeling: New Directions for Hydrologic Prediction
Publ.no.181 (1989), price $50

Regional Characterization of Water Quality
Publ.no.182 (1989), price $45

Snow Cover and Glacier Variations
Publ.no.183 (1989), price $30

Sediment and the Environment
Publ.no.184 (1989), price $40

Groundwater Contamination
Publ.no.185 (1989), price $40

Remote Sensing and Large-Scale Global Processes
Publ.no.186 (1989), price $40

FRIENDS in Hydrology. Proceedings of the Bolkesjø Symposium, April 1989
Publ.no.187 (1989), price $50

Groundwater Management: Quantity and Quality
Proceedings of the Benidorm Symposium, October 1989
Publ.no.188 (1989), price $60

Erosion, Transport and Deposition Processes. Proceedings of the Jerusalem Workshop, March-April 1987
Publ.no.189 (1990), price $40

Hydrology of Mountainous Areas. Proceedings of the Strbske Pleso Workshop, Czechoslovakia, June 1988
Publ.no.190 (1990), price $45

Regionalization in Hydrology. Proceedings of the Ljubljana Symposium, April 1990
Publ.no.191 (1990), price $45

First of New Series!

Hydrological Phenomena in Geosphere-Biosphere Interactions: Outlooks to Past, Present and Future
by Mălin Falkenmark
Monograph no.1 (1989), price $15

Available only from IAHS Press, Wallingford

PLEASE SEND ORDERS AND/OR ENQUIRIES TO:

Office of the Treasurer IAHS
2000 Florida Avenue NW
Washington, DC 20009, USA
[telephone: 202 462 6903]

Bureau des Publications de l'UGGI
140 Rue de Grenelle, 75700 Paris, France
[téléphone: 45 50 34 95 ext. 816; telex: 204989 igngnl f (Attn: UGGI)]

IAHS Press, Institute of Hydrology
Wallingford, Oxfordshire OX10 8BB, UK
[telephone: (0)491 38800; telex: 849365 hydrol g; fax: (0)491 32256]

Erosion, Transport and Deposition Processes

Edited by

D. E. WALLING

Department of Geography, University of Exeter, Amory Building, Rennes Drive, Exeter, Devon, EX4 4RJ, UK

A. YAIR & S. BERKOWICZ

Department of Physical Geography, Hebrew University of Jerusalem, Jerusalem, Israel 91904

Proceedings of a workshop held at Jerusalem, Israel, March–April 1987. The workshop was organized jointly by the International Commission on Continental Erosion of the International Association of Hydrological Sciences (IAHS) and the Commission on Measurement, Theory and Application in Geomorphology (COMTAG) of the International Geographical Union (IGU)

IAHS Publication No. 189

Published by the International Association of Hydrological Sciences 1990.

IAHS Press, Institute of Hydrology, Wallingford, Oxfordshire OX10 8BB, UK.

IAHS Publication No. 189.

ISBN 0-947571-37-X.

Thanks are due to everyone who has helped with the production of this volume. Several members of COMTAG and of ICCE kindly assisted with reviewing the manuscripts and the authors are to be particularly thanked for their patience in accepting the various delays that have accompanied the publication of their papers. Sarah Cage, from the IAHS Press in Wallingford, deserves special thanks for her work in producing the camera-ready copy from our edited manuscripts and for coordinating the printing and publication of the volume.

Des Walling,
Exeter

Aaron Yair & Simon Berkowicz,
Jerusalem

The camera-ready copy for the papers was prepared at IAHS Press, Wallingford, on an Advent desktop publishing system.

Printed in The Netherlands by Krips Repro Meppel.

Preface

The 13 papers presented in this volume represent one outcome of the International Workshop on Erosion, Transport and Deposition Processes with Particular Reference to Semi-arid and Arid Areas, held in Jerusalem - Beer Sheva - Elat in March-April 1987. Another set of contributions to this Workshop has been published as Catena Supplement 14, 1989.

The Workshop was a joint activity of the Commission on Measurement, Theory and Application in Geomorphology (COMTAG) of the International Geographical Union, and the International Commission on Continental Erosion (ICCE) of the International Association of Hydrological Sciences. It represents the first formal collaboration between these two groups of geomorphologists and hydrologists, although there have been close contacts in the past. The lively discussions in the lecture hall as well as in the field reflected a growing feeling among hydrologists and geomorphologists that, over the past two decades, computer modelling of runoff and erosion processes has been losing much of the necessary basis demonstrable by field relationships. Real world information is essential to enable us to understand natural processes scientifically; often, if possible, with the aid of computer modelling, but sometimes also without it. This order of precedence, irrespective of the methods used, is exemplified in the diverse papers contained in this volume which, we hope, will contribute to narrowing the gap between theory and modelling of erosion and sedimentation processes and their actual behaviour in the real world.

Semi-arid and arid terrains provide the field scientist with a mixed bag of advantages and disadvantages for research. Difficult logistics and an often short and erratic database are balanced by the relative simplicity of physiographic structure and minimal interference from antecedent events. Although considerable progress has been accomplished in recent years, as, for example demonstrated during the field excursions of this Workshop, the study of many important aspects of erosion processes in deserts is still in its infancy. Much more needs to be done in order to form a coherent and generalized body of knowledge applicable to severe environmental problems associated with erosion and sedimentation in arid areas, especially in developing countries. We hope that this Workshop has contributed towards this aim and that it will be followed by an even closer cooperation, in future, between hydrologists and geomorphologists, not only on the topic of desert processes but also in other fields.

Asher P Schick
Chairman, Commission on Measurement, Theory and Application in Geomorphology, International Geographical Union

Des Walling
President, International Commission on Continental Erosion, International Association of Hydrological Sciences

Contents

Preface *by Asher P. Schick & Des Walling* v

Field experiments on the resistance to overland flow of desert hillslopes *by Athol D. Abrahams, Anthony J. Parsons & Shiu-Hung Luk* 1

The sedimentary data base: an appraisal of lake and reservoir sediment based studies of sediment yield *by Ian D. L. Foster, John A. Dearing, Robert Grew & Karl Orend* 19

Empirical relationships for the transport capacity of overland flow *by Gerard Govers* 45

Evolution of an anthropogenic desert gully system *by Martin J. Haigh* 65

The relationship between sediment delivery ratio and stream order: a Romanian case study *by Ionită Ichim* 79

A simulation model for desert runoff and erosion *by Mike Kirkby* 87

Spatial variability of overland flow in a small arid basin *by Hanoch Lavee & Aaron Yair* 105

Towards a dynamic model of gully growth *by Anne C. Kemp* 121

The dynamics of gully head recession in a savanna environment *by Emmanuel Ajayi Olofin* 135

Conditions for the evacuation of rock fragments from cultivated upland areas during rainstorms *by J. W. A. Poesen* 145

Seasonal variations of runoff rates from field plots in the Federal Republic of Germany and in Hungary during dry years *by G. Richter & A. Kertész* 161

New developments in measuring bed load by the magnetic tracer technique *by Raymund Spieker & Peter Ergenzinger* 169

Some applications of caesium-137 in the study of erosion, transport and deposition *by D. E. Walling & S. B. Bradley* 179

Erosion, Transport and Deposition Processes (Proceedings of the Jerusalem Workshop, March-April 1987). IAHS Publ. no. 189, 1990.

Field experiments on the resistance to overland flow on desert hillslopes

ATHOL D. ABRAHAMS

Department of Geography, State University of New York at Buffalo, Buffalo, New York 14260, USA

ANTHONY J. PARSONS

Department of Geography, University of Keele, Keele, Staffordshire ST5 5BG, UK

SHIU-HUNG LUK

Department of Geography, Erindale Campus, University of Toronto, Mississauga, Ontario L5L 1C6, Canada

Abstract At-a-section and downslope variations in resistance to overland flow on desert hillslopes have been investigated by performing a series of experiments on six runoff plots in southern Arizona. The surfaces of these plots are irregular and covered with stones. As overland flow increases, the stones and microtopographic protuberances, which constitute the major roughness elements, are progressively inundated, thereby altering the flow resistance. Analyses of 14 cross sections reveal that the relation between the Darcy-Weisbach friction factor f and the Reynolds number Re has two basic shapes: convex-upward and negatively sloping. These shapes are explained in terms of the simultaneous operation of two processes. The first is the progressive inundation of roughness elements and increase in their wetted upstream projected area as discharge increases. This process causes f to increase. The second is the progressive increase in the depth of flow over already inundated parts of the bed as discharge increases. This process causes f to decline. Whether the f-Re relation has a positive or a negative slope depends on whether the first or the second process dominates, and this depends on the configuration of the bed and level of discharge. These findings have profound implications for the mathematical modelling of overland flow on desert hillslopes, as the computed overland flow hydrograph is very sensitive to the form of the f-Re relation. The downslope analyses indicate that there is a general tendency for f to decrease down the runoff plots, owing to the progressive downslope concentration of flow. This finding contrasts with Emmett's conclusion that f remains approximately constant downslope because of ponding behind microtopographic highs, notably vegetation mounds. Our finding appears to be more representative than Emmett's of sparsely vegetated desert hillslopes.

INTRODUCTION

In desert landscapes virtually all runoff from hillslopes occurs in the form of overland flow. On such hillslopes, overland flow generally appears as a shallow sheet of water with threads of deeper, faster flow diverging and converging around surface protuberances, rocks, and vegetation. As a result of these diverging and converging threads, flow depth and velocity vary markedly over short distances, giving rise to changes in the state of flow. Thus, over a small area the flow may be wholly laminar, wholly turbulent, wholly transitional, or consist of patches of any of these three flow states (Horton, 1945). Abrahams *et al.* (1986) termed this patchy kind of overland flow *composite* flow.

The resistance to overland flow offered by the surface of a hillslope may be expressed by a hydraulic roughness coefficient. The coefficient used in this study is the dimensionless Darcy-Weisbach friction factor f defined by

$$f = \frac{8gRs_e}{v^2} \quad (1)$$

where g is the acceleration due to gravity, R the hydraulic radius, S_e the energy slope, and v the mean flow velocity. Resistance to flow generally varies with rate of flow, and this variation is normally examined by plotting f against the dimensionless Reynolds number *Re* on log-log axes. The definition of *Re* employed here is

$$Re = 4vR/\upsilon \quad (2)$$

where υ is the kinematic fluid viscosity.

The relationship between f and *Re* is well established for pipes and smooth channels (Chow, 1959, p. 9-10), and laboratory experiments and theoretical analyses since the 1930s have shown that a similar relationship obtains for shallow flow over a plane bed with either a smooth or rough surface, provided that the surface is completely submerged by the flow (e.g. Horton *et al.*, 1934; Woo & Brater, 1961; Emmettt, 1970; Yoon & Wenzel, 1971; Shen & Li, 1973; Phelps, 1975; Savat, 1980). In these circumstances, the shape of the *f-Re* relation is a function of the state of flow: the relation has a slope of −1.0 where the flow is laminar and a slope close to −0.2 where the flow is turbulent (e.g., Horton *et al.*, 1934; Emmett, 1970; Morgali, 1970; Yoon & Wenzel, 1971; Shen & Li, 1973). The situation is less clear where the flow is transitional. In some experiments, the slope of the *f-Re* relation becomes positive in the transition zone (e.g., Phelps, 1975), whereas in others it does not (e.g., Yoon & Wenzel, 1971; Savat, 1980).

The form of the *f-Re* relation is of fundamental importance to the mathematical modelling of overland flow. Whether a model is based on the Saint Venant equations (Woolhiser & Liggett, 1967; Woolhiser, 1975) or employs the kinematic-wave approximation (e.g., Henderson & Wooding, 1964; Woolhiser & Liggett, 1967; Foster *et al.*, 1968; Lane & Woolhiser,

1977; Dunne & Dietrich, 1980; Moore, 1985), it needs to take into account the resistance to flow of the hillslope surface. In the past this usually has been done by incorporating into the model a relation between f and Re or surrogates thereof. The most widely used relation has been the conventional one for shallow flow over a plane bed. This relation generally has an adjustable parameter, such as the intercept of the laminar flow portion of the relation. In some cases the value of the intercept has been estimated from experimental data (e.g. Dunne & Dietrich, 1980), whereas in others it has been obtained by optimization (e.g. Woolhiser *et al.*, 1970; Woolhiser, 1975; Lane & Woolhiser, 1977). Studies have shown that the computed overland flow hydrograph is very sensitive to the value of the intercept (e.g. Woolhiser, 1975; Dunne & Dietrich, 1980), and as Fig. 1 shows, it is also extremely sensitive to the slope coefficient of the f-Re relation.

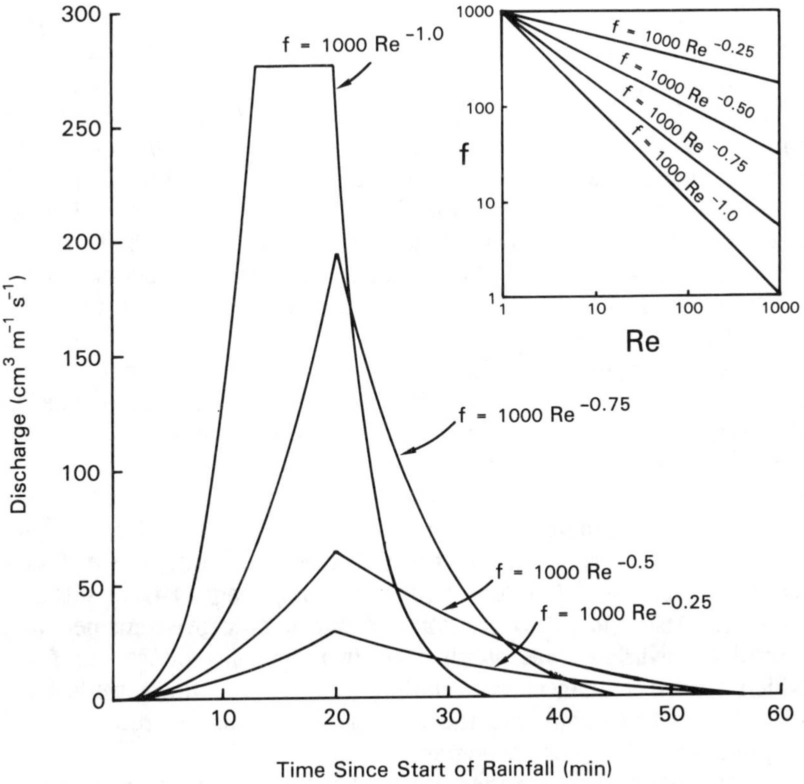

Fig. 1 Overland flow hydrographs predicted by the kinematic-wave model (Woolhiser, 1975) modified to allow for the effect of infiltration on the recession limb (Dunne & Dietrich, 1980). The figure shows the sensitivity of the hydrograph to the slope coefficient of the f-Re relation (inset) when all other factors remain constant, namely: slope length = 100 m, gradient = 0.09, rainfall duration = 20 min precipitation excess = 10 mm h^{-1}, infiltration capacity = 5 mm h^{-1}.

Although the conventional *f-Re* relation has been widely used in overland flow modelling, it is not at all clear that this relation applies to desert hillslopes, particularly as the irregular and stone-covered surfaces of such hillslopes induce composite overland flow. There has been no previous study for the *f-Re* relation on desert hillslopes. However, Roels (1984) computed the *f-Re* relations for 12 small (0.5 m^2) runoff plots with rilled and prerilled surfaces in the Ardeche drainage basin, France. Although *Re* ranged in value from 75 to 14 000 (with most observations being in the range $175 < Re < 7000$), none of the *f-Re* relations exhibited either a break or reversal in slope indicative of the transition from laminar to turbulent flow. Disregarding the relation for plot 12 which was on bare rock, the remaining 11 relations had slopes ranging from −0.07 to −0.90. Roels interpreted these negatively sloping *f-Re* relations as signifying that the flow was either transitional or turbulent. However, it is conceivable that the form of these relations reflected not so much the state of flow as of the character of the ground surface.

From Roels' descriptions of his plots it is evident that their surfaces were highly irregular and covered with stones that protruded above the flow. The microtopography and stones would not only have contributed to the flow resistance but would have given rise to composite flow. When composite flow prevails, a given *Re* value for a plot or cross section cannot be associated with a single state of flow. Thus, it is hardly surprising that Roels' *f-Re* relations depart from the conventional relation for shallow flow over a plane bed. Roels' study therefore casts doubt on the applicability of the conventional *f-Re* relation to desert hillslopes with irregular surfaces and points to the need for further field studies of the relation between *f* and *Re*.

Whereas Roels was concerned with the variation in flow resistance at a cross section, Emmett (1970) investigated the variation in this property down a hillslope. Using simulated rainfall to generate overland flow, he analysed the downslope variation in *f* with *Re* in seven hillslopes plots in semiarid Wyoming, USA. Emmett noted that flow resistance consists of both particle resistance (offered by sand grains and plant stems) and form resistance (caused by topographic irregularities), and he concluded that form resistance was the more important on the hillslopes he studied. He found that the slopes of the *f-Re* relations for his seven plots ranged from −0.80 to 1.00 and averaged 0.08. That flow resistance (on the average) remained approximately constant downslope was ascribed to two opposing effects on flow resistance which tended to cancel one another out. These effects were the downslope increase in flow depth and the downslope decrease in runoff efficiency owing to ponding behind microtopographic highs.

This paper presents the results of a field investigation of the variation in resistance to overland flow both at-a-section and downslope on desert hillslopes in southern Arizona. The at-a-section analyses examine the form of the *f-Re* relation and the factors that control this form, whereas the downslope analyses focus on whether *f* increases, decreases, or remains constant downslope and the reasons for this behaviour. The findings of both sets of analyses deviate in important ways from those of previous studies and have significant implications for runoff and erosion on desert hillslopes.

STUDY AREA, METHODS, AND DATA

Six runoff plots, hereafter referred to as P1, ..., P6, were established on hillslopes ranging in gradient from 6° to 33° in the Walnut Gulch Experimental Watershed, Tombstone, Arizona (31° 43'N, 110° 41'W). The plots, which were approximately 5.5 m long and 1.8 m wide, were underlain by Bascal gravelly fine sandy loam (W. E. Emmerich, personal communication, 1986) developed on weakly consolidated, coarse Quaternary alluvium. The surfaces of the plots were largely covered by gravel that had weathered out of the underlying material. The vegetation on the hillslopes is dominated by desert shrubs, notably sandpaper bush *(Mortionia scabrella)* and creosote bush *(Larrea tridentata)*. However, the runoff plots were located between shrubs and had only a sparse cover of forbs and grasses that were clipped 1-2 cm above the ground for the runoff experiments. The locality has a warm, semiarid climate with a mean annual precipitation of 288 mm and a mean monthly temperature range of 8° to 27°C (Osborn, 1983).

Each plot was equipped with separate sprinkle and trickle systems to simulate rainfall and overland flow, respectively. The design and operation of these systems have been described elsewhere (Luk *et al.*, 1986). Eight experimental runs, designated R1, ..., R8, were performed on each plot. During the first run, rainfall alone was applied to each plot at a rate of 145 mm h^{-1} for approximately 40 minutes. A natural rainstorm of this intensity and duration has a recurrence interval in excess of 200 years at Walnut Gulch (Osborn, 1983). A rainfall intensity of 145 mm h^{-1} was selected because trial and error revealed that an intensity of this order was necessary to generate Horton overland flow. Even then, overland flow during the first run on P5 and P6 was so shallow that no depth measurements were attempted. During the second run, rainfall at a rate of 72 mm h^{-1} for approximately 30 minutes was combined with overland flow, which was trickled onto the upper part of each plot at a rate of 572 cm^3 s^{-1}. A natural rainstorm of this intensity and duration has a recurrence interval of about 10 years at Walnut Gulch (Osborn, 1983). During the remaining six runs, overland flow alone was applied to each plot at inflow rates of 572, 733, 908, 1067, 1233, and 1400 cm^3 s^{-1} for R3 to R8, respectively. The only exception was P1-R3 where the inflow rate was 533 cm^3 s^{-1}.

The outflow rate for each run was measured volumetrically by directing the flow at the lower end of the plot into a collecting trough and periodically measuring the time required to almost fill a 2 litre bottle. The weight of the fluid in each bottle was subsequently measured in the laboratory, and its volume determined by assuming a density of 1 g cm^{-3}.

Three cross sections, henceforth identified as C1, C2, and C3, were established on each plot at 1.25, 3.25, and 5.25 m from the top of the plot. Discharges Q at these cross sections were computed by assuming that infiltration and evaporation losses were uniformly distributed over the plot. During all but a few runs, which are identified below, the outflow rate eventually became more or less constant. When this occurred, the depth of flow to the nearest millimetre was measured with a thin scale at 5 cm intervals across each cross section, and these measurements were averaged to yield the mean flow depth d. The mean velocity (v) was calculated by dividing Q by $w.d$, where w is the width of the cross section. Owing to the stony and uneven nature of the ground surface, a combination of sheet-metal strips and cement walls was used to delimit the runoff plots. As a result, the cross sections vary somewhat in width.

As in all previous field studies of overland flow, d is employed in place of R, and S_e is approximated by local ground slope S. The value of S for each cross section was determined by laying a 0.3 m long board along the flow lines of the two or three major threads of flow, measuring the gradient of the board in each case, and averaging these gradients (Table 1). υ was estimated from water temperature, and f and Re were calculated using equations (1) and (2).

The analyses undertaken in this study are described in the following two sections. The analyses in the first section are concerned with the at-a-section variation in f and are based on R3 through R8. R1 and R2 were excluded because they involve rainfall, which is an additional factor influencing resistance to flow (Yoon & Wenzel, 1971; Shen & Li, 1973). The analyses in the second section focus on the downslope variation in f. Data from all eight runs are analysed, but the analyses of R1 and R2 are performed separately from those of R3 through R8.

AT-A-SECTION ANALYSES

Depth versus discharge

Emmett (1970) demonstrated that the hydraulic geometry approach to river channels pioneered by Leopold & Maddock (1953) could also be employed to analyse the hydraulics of overland flow. The simple log-linear version of the hydraulic geometry relations used by Leopold and Maddock was generalized to include curvilinear trends by Richards (1973) who utilized log-quadratic functions. In this study we are concerned only with the relation between d and Q. The log-quadratic form of this relation is

$$\log d = f_1 + f_2 \log Q + f_3 (\log Q)^2 \qquad (3)$$

For each of the three cross sections on the six plots, d was plotted against Q on log-log graph paper (Fig. 2), and a quadratic equation of the form of equation (3) was fitted by least squares to each scatter of points, with the quadratic term being included only if its contribution to the explained variance was significant at the 0.10 level. Inasmuch as the variance accounted for by the fitted equations always exceeds 90%, the equations closely approximate their equivalent structural relations (Mark & Church, 1977).

Friction factor versus Reynolds number

Four of the 18 cross sections, namely P2-C3, P3-C3, P6-C2, and P6-C3, were

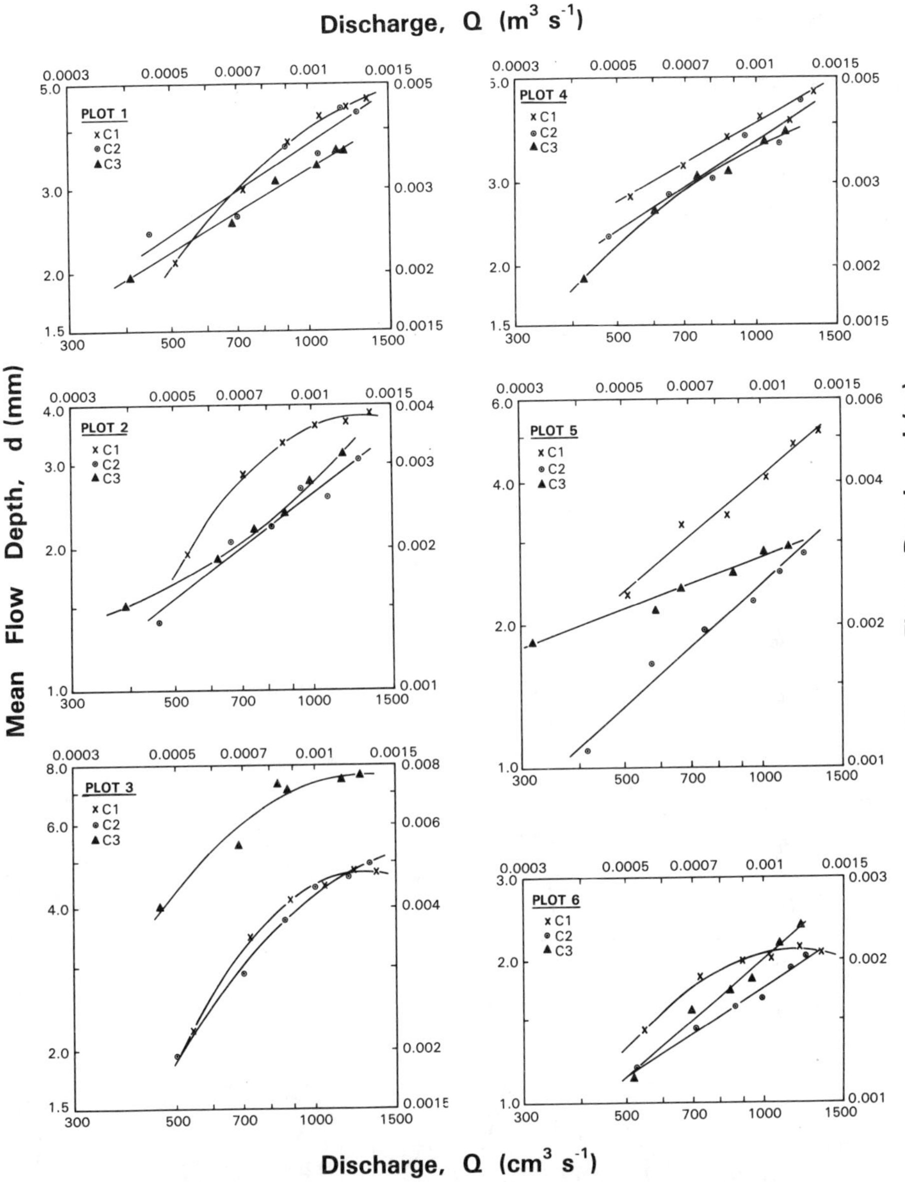

Fig. 2 Graphs of mean flow depth against discharge.

discarded from the analyses because they were affected by erosion of a plunge pool beneath a rock step, formation of an organic dam, or scouring by debris flows. For the remaining 14 cross section *f* was plotted against *Re* on log-log axes (Fig. 3). The relation between *f* and *Re* for each cross section was then determined algebraically from the corresponding *d-Q* relation (Abrahams *et al.*, 1986). Because the general form of the *d-Q* relation is log-quadratic, so is the general form of the *f-Re* relation:

$$\log f = k_1 + k_2 \log Re + k_3 (\log Re)^2 \quad (4)$$

The computed *f-Re* relations for the 14 cross sections are graphed in Fig. 3. Six of these relations are convex-upward with vertices within the observed range of *Re*. The remaining eight relations are linear, three being positively sloping and five negatively sloping. The range of k_2 values for the negatively sloping relations is −0.078 to −0.893, which is similar to that reported by Roels (1984).

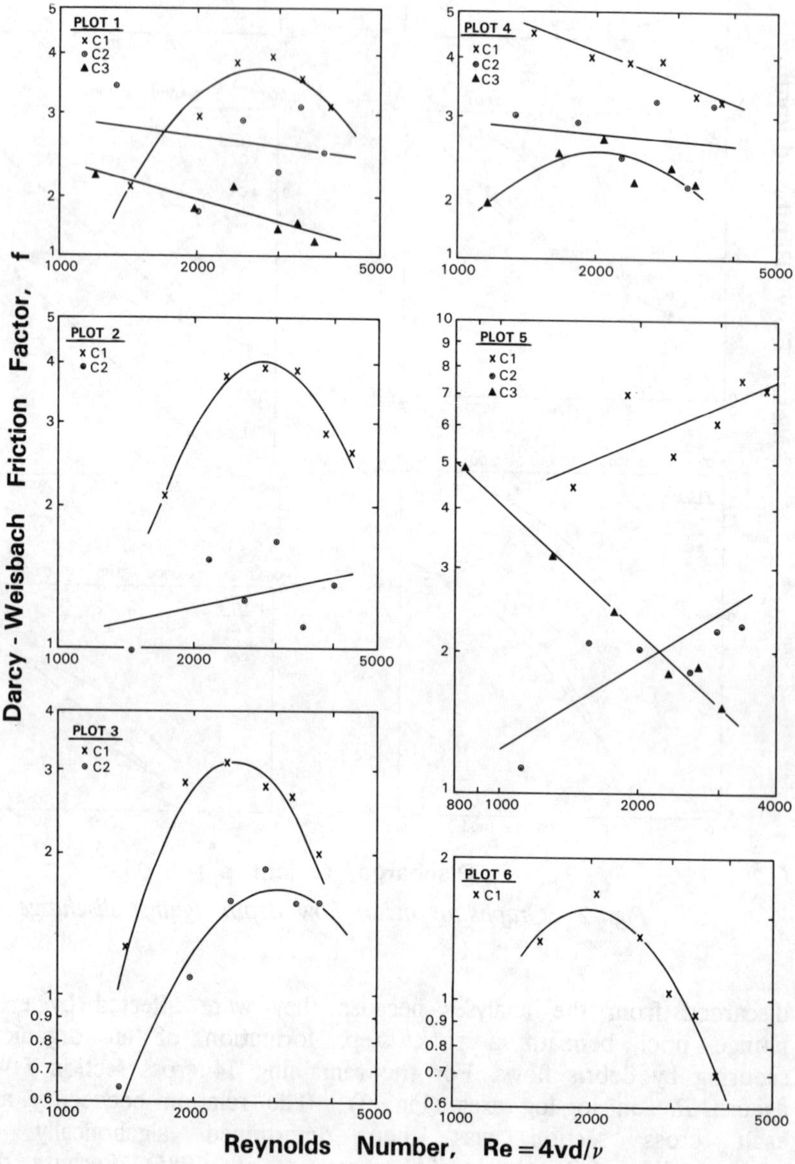

Fig. 3 Graphs of Darcy-Weisbach friction factor against Reynolds number.

Inundation of roughness elements

The shapes of the *f-Re* relations in Fig. 3 bear no resemblance to the conventional *f-Re* relation for shallow flow over a plane bed. Thus it is unlikely that they can be explained in terms of changes in the state of flow, and some other explanations must be sought. Abrahams *et al.* (1986) proposed that these shapes can be understood in terms of the simultaneous operation of two processes. As discharge and hence *Re* increase (a) the flow progressively inundates the protruding roughness elements, thereby increasing the wetted upstream projected area of these elements and the resistance to flow (Herbich & Shulits, 1964); and (b) the depth of flow over already inundated portions of the cross section increases, thereby increasing relative submergence (i.e., the ratio of flow depth to height of the drowned roughness elements) and causing flow resistance to decline (Bathurst, 1985). Whether the *f-Re* relation has a positive or negative slope at a particular value of *Re* depends on whether the first or the second process dominates.

On many, if not most, desert hillslopes (depicted schematically in Fig. 4) the *f-Re* relation is convex-upward because when discharge is low but increasing, the flow progressively inundates roughness elements, causing the wetted upstream projected area of the elements to increase rapidly, and *f* to increase. However, a point is soon reached when most of the elements are largely or wholly submerged, and any further increase in discharge is

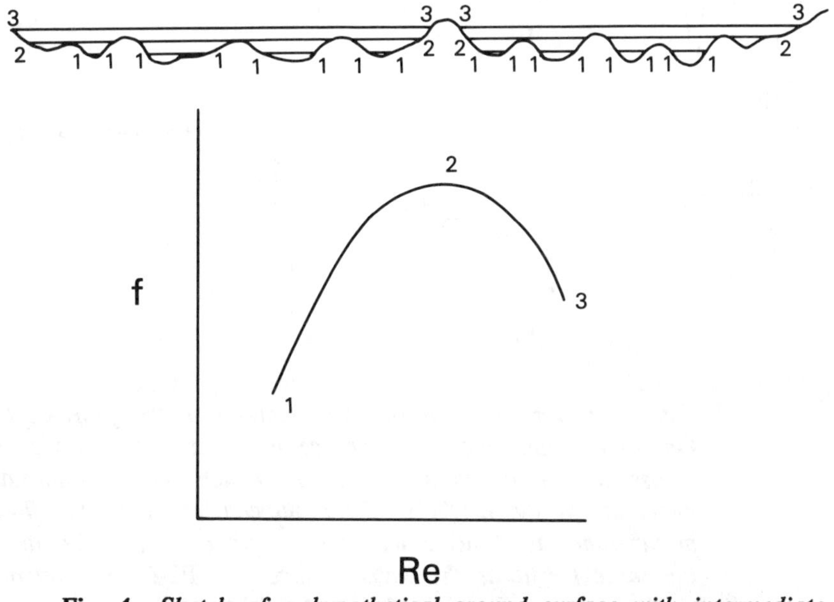

Fig. 4 Sketch of a hypothetical ground surface with intermediate-sized roughness elements of varying height showing how the progressive inundation of such a surface gives rise to a convex-upward f-Re relation. The numbers identify particular water surface levels and their corresponding f and Re values.

accompanied by a progressively slower rate of increase in wetted upstream projected area of the elements. As discharge increases, so does depth of flow. Eventually the effect of this process outweighs that of increases in wetted upstream projected area of the roughness elements, and *f* begins to decrease. An example is provide by P2-C1 (Fig. 5(a)).

Negatively sloping *f-Re* relations appear to be produced by two types of bed configurations (shown schematically in Fig. 6). The first is an almost planar bed with many small roughness elements and perhaps the occasional, widely spaced, large element. Even at low flow virtually all the small elements are submerged, and *f* is relatively large. As discharge increases, the wetted upstream projected area of the roughness elements increased only slowly because all the elements save the largest are already submerged. Meanwhile, most of the bed becomes drowned under greater depths of flow, causing *f* to decline. P4-C1 is an example of this bed configuration (Fig. 5(b)).

The second type of bed configuration is characterized by pronounced concentrations of flow; such as in rills. Roels' (1984) field experiments were conducted on this kind of surface. As discharge increases, the flow remains confined within the rills, and depth of flow within each rill increases rapidly.

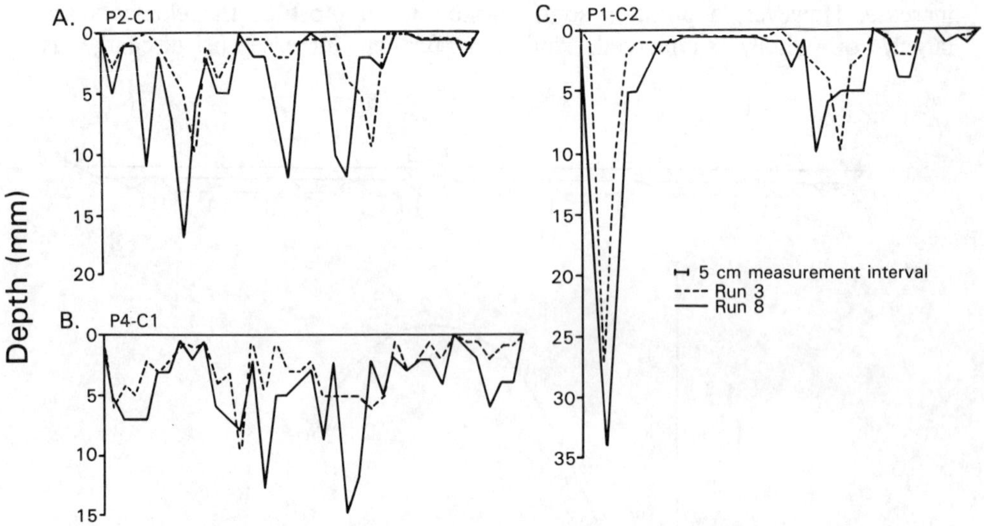

Fig. 5 Graphs of selected cross sections drawn from depth data. On these graphs the water surface is depicted as a horizontal line across the entire cross section. Inasmuch as no information is available on the height of the protuberances above the flow, each protuberance is depicted as either a point or a horizontal line at the same height as the water surface. (a) P2-C1 has intermediate-sized roughness elements of varying height, the progressive inundation of which produces a convex-upward f-Re relation. (b) P4-C1 is almost planar with multiple small roughness elements, which become drowned at low flow and give rise to a negatively sloping f-Re relation. (c) P1-C2 is rilled and has a negatively sloping f-Re relation.

Thus, each rill becomes hydraulically more efficient, and the f value for the cross section decreases. P1-C2 is an example (Fig. 5(c)).

For some of the cross sections on our plots positively sloping *f-Re* relations were obtained. These relations appear to be the left-hand limbs of convex-upward relations whose vertices lay beyond the observed range of *Re*. The relatively rapid increase in the wetted upstream projected area of the roughness elements as the flow inundates the cross section and causes f to increase cannot continue indefinitely as *Re* increases, so the slope of the *f-Re* relation must eventually become negative.

Thus it is concluded that *f-Re* relations for desert hillslopes have two basic shapes: convex-upward and negatively sloping. These shapes arise from the progressive inundation of ground surfaces with different configurations. As overland flow on desert hillslopes is composite, changes in state of flow are difficult to define and appear to have relatively little influence on the shape of the *f-Re* relation. Although some progress has been made in linking the shape of the relation to the configuration of the ground surface, the linkage is obviously complex and requires a great deal more study.

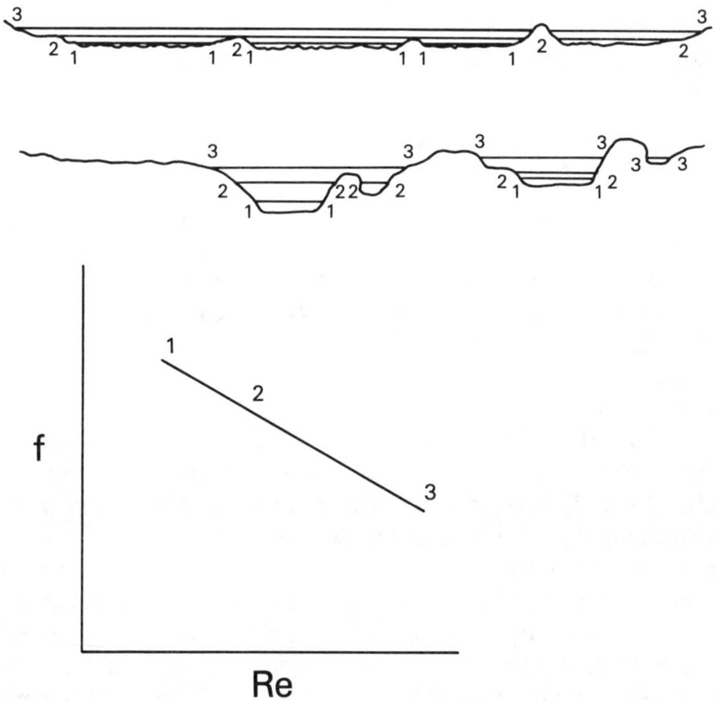

Fig. 6 Sketches of two hypothetical ground surfaces, the first being almost planar with multiple small roughness elements and the second being rilled, showing how the progressive inundation of such surfaces gives rise to negatively sloping f-Re relations. The numbers identify particular water surface levels and their corresponding f and Re values.

DOWNSLOPE ANALYSES

Overland flow without rainfall

The at-a-section analyses were concerned with the variation in f with respect to Re, where $Re = 4Q/w\upsilon$ and w and υ are constant. However, Re cannot be used in the analysis of downslope changes in resistance to flow because, although Q always decreases systematically downslope, Re does not. This is because irregularities in the walls of the runoff plots cause w to vary between cross sections. Indeed, during some runs w actually decreases downslope from one cross section to the next by a larger proportion than Q decreases, causing Re to increase. Consequently, Re does not always decrease downslope and is an unreliable indicator of the downslope direction. Therefore in the downslope analyses f was analysed with respect to Q rather than Re. Furthermore, because of the variation in w, the downslope relations between f and Q could not be computed algebraically from those between d and Q, as in the at-a-section analyses. Accordingly, the relations between f and Q were determined empirically.

Each downslope analysis examined the variation in f as Q declined from C1 to C3 during a single run on a given plot. Because there were only three cross sections, the relation between f and Q had to be computed from a maximum of three points. So we chose to represent it by a simple log-linear equation of the form.

$$\log f = j_1 + j_2 \log Q \qquad (5)$$

Inasmuch as P6-C2 and P6-C3 had been earlier discarded, only one cross section remained on P6, and it was impossible to compute any f-Q relations for this plot. P2-C3 and P3-C3 had also been discarded, leaving two cross sections each on P2 and P3. The f-Q relations for these plots were therefore calculated from just two points. For the remaining three plots the structural relation between f and Q for each run was estimated by computing the reduced major axis (Miller & Kahn, 1962, p. 204-205).

Downslope f-Q relations were determined for a total of 30 runs on five plots. For 28 of the relations, the computed correlation coefficients exceed 0.60 (Table 1), signifying at least modest linearity in the data. The correlation coefficients for the remaining two relations are −0.10 and −0.13, denoting essentially no linearity. These two relations were therefore disregarded. The computed values of j_2 for the remaining 28 relations range from 2.39 to 62.71 and have a mean of 11.05 (Fig. 7). Thus all the values of j_2 are positive, indicating that when there is a systematic change in f as discharge decreases down the runoff plots; it always decreases. When discharge increases downslope under natural rainfall, resistance to flow might be expected to decrease owing to an increase in relative submergence (Emmett, 1970). However, inasmuch as discharge always decreases down the runoff plots during R3 through R8, the decrease in f cannot be explained in this way, and one must seek another explanation.

The decrease in f appears to be due to the progressive concentration of flow

Table 1 Slope and correlation coefficients for downslope friction factor-discharge relations

Plot	Run	Downstream change in discharge*	Slope coefficient (j_2)	Correlation coefficient
P1	R1	-	--**	--
	R2	D	70.57	0.73
	R3	D	-0.23	-0.10
	R4	D	7.68	0.85
	R5	D	9.33	0.97
	R6	D	62.71	0.99
	R7	D	12.41	0.94
	R8	D	5.46	0.99
P2	R1	-	--**	--
	R2	D	10.76	1.00***
	R3	D	5.17	1.00
	R4	D	14.15	1.00
	R5	D	15.12	1.00
	R6	D	10.53	1.00
	R7	D	9.73	1.00
	R8	D	9.34	1.00
P3	R1	I	-0.28	1.00***
	R2	I	-151.05	1.00
	R3	D	8.70	1.00
	R4	D	35.90	1.00
	R5	D	20.85	1.00
	R6	D	8.04	1.00
	R7	D	19.63	1.00
	R8	D	6.00	1.00
P4	R1	-	--**	--
	R2	D	32.68	0.83
	R3	D	3.51	1.00
	R4	D	3.03	0.98
	R5	D	2.50	0.75
	R6	D	3.76	0.99
	R7	D	2.54	0.75
	R8	D	2.65	0.89
P5	R1	-	--****	--
	R2	D	4.92	0.78
	R3	D	-0.44	-0.13
	R4	D	2.39	0.61
	R5	D	3.26	0.76
	R6	D	6.35	0.81
	R7	D	8.94	0.91
	R8	D	9.62	0.96

* D = decrease; I = increase
** Constant discharge not attained.
*** The perfect correlations for the runs on this plot are due to the relations being fitted to only two points.
**** Overland flow was too shallow for accurate depth measurements to be made.

down the runoff plots. On P1, P2, P3 and P5 this concentration of flow is aided by the plot walls. On these plots some of the overland flow is directed by the microtopography toward one of the side walls where it forms a major thread that continues downslope adjacent to the wall. However, even where the flow is not confined by the side walls, it tends to concentrate downslope, as on P4. Thus the field experiments suggest that on the desert hillslopes under study, f decreases downslope owing to a tendency for overland flow to concentrate in that direction.

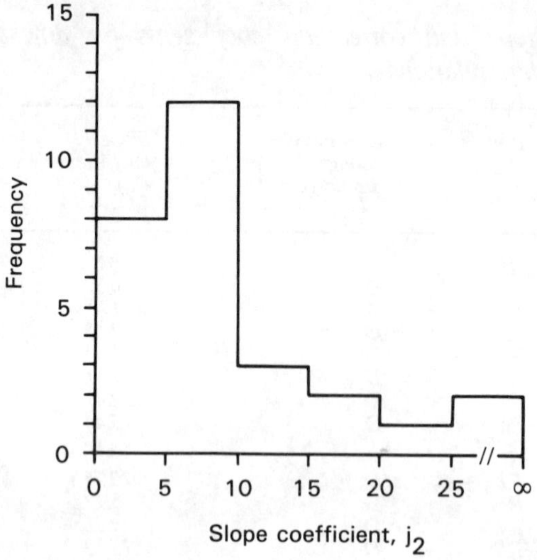

Fig. 7 Frequency distribution of the slope coefficient in the downslope f-Q relation for runs without rainfall.

Overland flow with rainfall

There is some question, however, whether these field experiments using overland flow without rainfall accurately simulate overland flow during natural rainstorms. In a recent study, Dunne & Aubry (1986) have argued that when overland flow occurs in the absence of rainfall, it is inherently unstable and incises rills. On the other hand, when it is accompanied by rainfall, rainsplash diffuses sediment from protuberances into rills and eradicates these features. Thus the possibility must be considered that the progressive concentration of overland flow down the runoff plots during R3 through R8 was the result of incision in the absence of rainfall and does not represent the actual situation in these desert hillslopes under natural rainfall. If overland flow does not concentrate downslope under rainfall, then f may not decrease downslope.

To investigate this possibility the downslope f-Q relations were analysed for the first two runs on each plot. During R1 simulated rainfall alone was applied to each plot, and during R2 rainfall was combined with inflow trickled onto the upper end of each plot. Certain runs, however, had to be excluded from the analysis. Specifically, P5-R1 and P6-R1 were excluded because the overland flow was so shallow that accurate depth measurements could not be made. P1-R1, P2-R1, and P4-R1 were omitted because during these runs the outflow continued to increase after measurements began. Thus steady state conditions, which are essential for any downslope analysis, did not obtain. Finally, P6-R2 was discarded because severe erosion during the run significantly altered the shapes of C2 and C3. Six runs therefore remained. However, the analyses of three of these runs, namely P2-R2, P3-R1, and P3-R2, were limited to two cross sections, as the mean flow depths were inflated by the scouring of a plunge pool at P2-C3 and the impounding of water behind an organic dam at P3-C3.

During two of these six runs, P3-R1 and P3-R2, discharge increased downslope, whereas during the other four runs, P1-R2, P2-R2, P4-R2, and P5-R2, it decreased. Table 1 shows that where discharge increased downslope j_2 is negative, and where discharge decreased downslope it is positive. In other words, f decreased down the runoff plots during all six runs. That f should decrease, despite a downslope decrease in discharge, during four of the runs implies that the decrease in f was the consequence of an increase in hydraulic efficiency associated with the progressive downslope concentration of flow. An examination of the cross sections for the six runs reveals that this concentration of flow was aided by a side wall during only one run. These results are consistent with those for overland flow without rainfall and leave little doubt that f decreases down the desert hillslopes under study during natural rainfall, and that this decrease is due to the progressive downslope concentration of flow.

This conclusion contrasts with Emmett's (1970) finding that f remains approximately constant down the semiarid hillslopes he investigated in Wyoming. Emmett attributed the fact that f does not decrease down his hillslopes to ponding of the flow behind microtopographic highs, notably vegetation mounds. There were few vegetation mounds on our runoff plots, and the only ponding we observed was behind occasional organic dams or in rare plunge pools. Thus the discrepancy between Emmett's and our findings appears to be due to differences in the relative importance of ponding, and this seems to be related to the density of vegetation mounds. On other hillslopes different factors may be important, so care must be taken in generalizing from either Emmett's or our findings. Even so, it seems likely that in desert landscapes where vegetation is sparse and ponding is rare, our findings will apply more widely than Emmett's, and overland flow will concentrate downslope, causing resistance to flow to decline.

SUMMARY AND CONCLUSION

This paper describes a series of experiments performed on six runoff plots in semiarid southern Arizona aimed at investigating both at-a-section and downslope variations in resistance to overland flow on desert hillslopes. The surfaces of these plots were irregular and covered with stones. As overland flow increased, the stones and microtopographic protuberances, which constituted the major roughness elements, were progressively inundated, thereby altering the flow resistance.

Analyses of 14 cross sections on the runoff plots suggest that at-a-station *f-Re* relations have two basic shapes: convex-upward and negatively sloping. These shapes can be understood in terms of the simultaneous operation of two processes. The first is the progressive inundation of roughness elements and increase in their wetted upstream projected area as discharge increases. This process causes flow resistance to increase. The second is the progressive increase in depth of flow over already inundated parts of the cross section as discharge increases. This process causes flow resistance to decline. On some hillslopes the first process dominates at low discharges and the second at high discharges to produce a convex-upward *f-Re* relation. However, on other hillslopes the first process never dominates, and the *f-Re* relation is negatively sloping over the full range of *Re* experienced. The present analyses indicate that this can occur where the flow remains concentrated as discharge increases or where the roughness elements are sufficiently small that they are rapidly submerged at low discharges.

This study therefore suggests that the conventional *f-Re* relation for shallow flow over a plane bed does not apply to desert hillslopes and should not be employed in mathematical models of overland flow on such hillslopes. In view of the sensitivity of the computed hydrographs to the *f-Re* relation, there is an urgent need to develop a practical and reliable method of estimating this relation for a given hillslope. The apparent dependence of the *f-Re* relation on the configuration of the surface suggests that the development of such a method will require a closer study of the connection between the relation and surface morphology. Given the irregular character of the ground surface and the composite nature of the flow on desert hillslopes, there is great scope for using laboratory experiments for this purpose. However, these experiments will need to be performed on realistic irregular surfaces, rather than on the planar ones employed heretofore.

The downslope analyses of *f* both with and without rainfall, indicate that there is a general tendency for *f* to decrease down the runoff plots owing to the progressive downslope concentration of flow and consequent increase in hydraulic efficiency. *f* decreases down the plots despite the fact that during almost every run discharge decreases downslope. Thus one might expect that during major natural storms when discharge increases downslope, the decrease in *f* would be even more pronounced.

This finding contrasts with Emmett's (1970) conclusion that *f* remains approximately constant downslope. The discrepancy appears to be due to the greater incidence on Emmett's hillslopes of ponding behind microtopographic highs, notably vegetation mounds. It therefore seems likely that our findings have wider applicability than Emmett's on desert hillslopes with sparse vegetation covers.

It remains unclear why there is a strong tendency on the hillslopes under study for overland flow to concentrate downslope, and further research is required on this subject. Whatever the reason, it appears certain that this tendency promotes the development of rills, which give rise to higher erosion rates than unconcentrated sheetflow (Kirkby & Kirkby, 1974; Meyer *et al.*, (1975). Thus the high drainage densities and sediment yields typically associated with semiarid environments may stem from this tendency. Further studies of downslope variations in flow resistance are therefore required. However, these need to be conducted at a larger scale than the runoff plots investigated here, ideally at the scale of the entire hillslope.

Acknowledgements This study was conducted in the Walnut Gulch Experimental Watershed, Tombstone, Arizona with the support and cooperation of the Agricultural Research Service, US Department of Agriculture. We are particularly grateful to Kenneth G. Renard for permission to use the accommodation and laboratory facilities at the Tombstone field station; to J. Roger Simanton for help and advice in initiating the research project; and to Howard Larsen, Raymond Brown, and James Smith for their assistance with the field work and laboratory analysis. The project was supported by NATO Grant RG. 85/0066 for International Cooperation in Research. Additional funding was provided to Luk by the Natural Sciences and Engineering Research Council of Canada.

REFERENCES

Abrahams, A. D., Parsons, A. J. & Luk, S. H. (1986) Resistance to overland flow on desert hillslopes. *J. Hydrol.* **88**, 343-363.

Bathurst, J. C. (1985) Flow resistance estimation in mountain rivers. *J. Hydraul. Div. ASCE* **111**, 625-643.

Chow, V. T. (1959) *Open-channel Hydraulics.* McGraw-Hill, New York.

Dunne, T. & Aubry, B. F. (1986) Evaluation of Horton's theory of sheetwash and rill erosion on the basis of field experiments. In: *Hillslope Processes* (ed. A. D. Abrahams), 31-53. Allen & Unwin, Boston, USA.

Dunne, T. & Dietrich, W. E. (1980) Experimental study of Horton overland flow on tropical hillslopes. 2. Hydraulic characteristics and hillslope hydrographs. *Z. Geomorphol. Suppl. Bd* **35**, 60-80.

Emmett, W. W. (1970) The hydraulics of overland flow on hillslopes. *USGS Prof. Pap.* 662-A.

Foster, G. R., Huggins, L. F. & Meyer, L. D. (1968) Simulation of overland flow on short field plots. *Wat. Resour. Res.* **4**, 1179-1187.

Henderson, F. M. & Wooding, R. A. (1964) Overland flow and groundwater flow from a steady rainfall of finite duration. *J. Geophys. Res.* **69**, 1531-1540.

Herbich, J. B. & Shulits, S. (1964) Large-scale roughness in open-channel flow. *J. Hydraul. Div. ASCE* **90**, 203-230.

Horton, R. E. (1945) Erosional development of streams and their drainage basins; hydrophysical approach to quantitative morphology. *Bull. Geol. Soc. Am.* **56**, 275-370.

Horton, R. E., Leach, H. R. & Van Cliet, R. (1934) Laminar sheet flow. *Trans. AGU* Part **2**, 393-404.

Kirkby, A. & Kirkby, M. J. (1974) Surface wash at the semi-arid break in slope. *Z. Geomorphol. Suppl.* Bd **21**, 151-176.

Lane, L. J. & Woolhiser, D. A. (1977) Simplification of watershed geometry affecting simulation of surface runoff. *J. Hydrol.* **35**, 173-190.

Leopold, L. B. & Maddock, T. (1953) The hydraulic geometry of stream channels and some physiographic implications. *USGS Prof. Pap.* 252.

Luk, S. H., Abrahams, A. D. & Parsons, A. J. (1986) A simple rainfall simulator and trickle system for hydro-geomorphological experiments. *Phys. Geogr.* **7**, 344-356.

Mark, D. M. & Church, M. (1977) On the misuse of regression in earth science. *Math. Geol.* **9**, 63-75.

Meyer, L. D., Foster, G. R. & Romkens, M. J. M. (1975) Source of soil eroded by water from upland slopes. In: *Present and Prospective Technology for Predicting Sediment Yields and Sources (Proceedings of the Sediment-Yield Workshop)* 177-189. USDA Sedimentation Laboratory, Oxford, Mississippi, USA.

Miller, R. L. & Kahn, J. S. (1962) *Statistical Analysis in the Geological Sciences.* Wiley, New York, USA

Moore, I. D. (1985) Kinematic overland flow: generalization of Rose's approximate solution. *J. Hydrol.* **82**, 233-245.

Morgali, J. R. (1970) Laminar and turbulent flow hydrographs. *J. Hydraul. Div.* ASCE **96**, 441-460.

Osborn, H. B. (1983) Precipitation characteristics affecting hydrologic response of southwestern rangelands. *Agriculture Reviews and Manuals, Western Series,* **34**, USDA, Oakland, California, USA.

Phelps, H. O. (1975) Shallow laminar flows over rough granular surfaces. *J. Hydraul. Div. ASCE* **101**, 367-384.

Richards, K. S. (1973) Hydraulic geometry and channel roughness - a non-linear system. *Am. J. Sci.* **273**, 877-896.

Roels, J. M. (1984) Flow resistance in concentrated overland flow on rough slope surfaces. *Earth Surf. Processes Landforms* **9**, 541-551.

Savat, J. (1980) Resistance to flow in rough supercritical sheet flow. *Earth Surf. Processes* **5**, 103-122.

Shen, H. W. & Li, R. M. (1973) Rainfall effect on sheet flow over smooth surface. *J. Hydraul. Div. ASCE* **99**, 771-792.

Woo, D. C. & Brater. E. F. (1961) Laminar flow in rough rectangular channels. *J. Geophys. Res.* **66**, 4207-4217.

Woolhiser, D. A. (1975) Simulation of unsteady overland flow. In: *Unsteady Flow in Open Channels*, Vol. II (eds. K. Mahmood & V. Yevjevich) 485-508. Water Resources Publications, Fort Collins, Colorado, USA.

Woolhiser, D. A. & Liggett, J. A. (1967) Unsteady, one-dimensional flow. *Wat. Resour. Res.* **3**, 753-771.

Woolhiser, D. A., Hanson, C. L. & Kuhlman, A. R. (1970) Overland flow on rangeland watersheds. *J. Hydrol.* (NZ) **9**, 336-356.

Yoon, Y. N. & Wenzel, H. G. (1971) Mechanics of sheet flow under simulated rainfall. *J. Hydraul. Div. ASCE* **97**, 1367-1386.

Erosion, Transport and Deposition Processes (Proceedings of the Jerusalem Workshop, March-April 1987). IAHS Publ. no. 189, 1990.

The sedimentary data base: an appraisal of lake and reservoir sediment based studies of sediment yield

IAN D. L. FOSTER, JOHN A. DEARING, ROBERT GREW, & KARL OREND

Centre for Environmental Science Research and Consultancy, Department of Geography, Coventry Polytechnic, Priory Street, Coventry, CV1 5FB, UK

Abstract Direct monitoring provides a highly resolved data base with which to study short term process response systems, whilst sedimentary information, based for example on stratigraphic and palaeoecological studies of river terraces and valley deposits, provide palaeohydrological information covering much of the Holocene period. Medium term changes, over decades and centuries, are much less well represented in the hydrological literature for two major reasons. First, monitoring programmes have not been established for a sufficient period of time to detail changes over these timescales and secondly, sedimentary features in river valleys are rarely deposited continuously over these timescales. These sediments do not preserve material of use for highly resolved dating and the relationships between sediment yield and deposition in such situations is complex. One solution to this problem is to analyse the accumulating sediments in natural lakes and reservoirs in order to estimate sediment yield changes through time. This paper reviews the application of the lake-watershed ecosystem framework over the last decade, in studies reconstructing sediment yield histories in a wide variety of geographical environments from northwest Europe, Scandinavia, North America, North Africa and Asia. An attempt is made to identify some of the technical problems in sampling lake sediments, in methods of dating and core correlation and in the interpretation of the sediment yield information, once collected, in order to account for non-catchment contributions to the sedimentary record. The evidence obtained to date shows that lakes in most environments preserve a record sufficiently well for the reconstruction of quantitative sediment yields. Geomorphological implications of the development of lake catchment studies are highlighted.

INTRODUCTION

In a recent UNESCO report for the International Hydrological Programme, Hadley (1985) has considered the monitoring requirements for the investigation of natural and manmade changes to the hydrological regime and related ecological environments. Within this brief report, six major objectives are identified for monitoring networks. These objectives are summarized below:

(a) to aid in the understanding of natural processes and how process rates are influenced by human activity and climate;
(b) to detect trends and periodicities in hydrologic time series;
(c) to formulate hypotheses for possible causes of change in the system;
(d) to measure process rates under natural and altered conditions;
(e) to describe qualitatively and measure quantitatively the effects of human activity on the environment; and
(f) to aid prediction in the management of land and water resources in the future.

Undoubtedly, process monitoring is essential for the collection of highly resolved data, the detailed examination of process dynamics and the construction of dynamic models of soil erosion and sediment transport in the fluvial environment. Direct process monitoring alone may, however, be inappropriate for the aims encompassed in points (b), (c), (e) and (f) above, because of the relatively short time scales concerned (less than a decade). The most obvious limitations here relate to the inherent variability in natural hydrological systems and the need to identify the time scales over which adjustments to human and natural changes occur. Streamflow and sediment and solute transport rates are inherently variable in contemporary systems. For example, Walling & Webb (1981) report the annual sediment yields for a 7 year continuously monitored record from the River Creedy, Devon UK. This record has an annual coefficient of variation of 57.9%. The record of sediment yields entering the Aswan High Dam for a 15 year period after completion have a coefficient of variation of 32.9% (based on data in Shalash, 1982). Furthermore, Foster (1980) has highlighted the importance of extreme events in runoff and solute load data for an East Devon catchment, where the annual coefficient of variation for a 3 year record including the 1976 UK drought was 63.9% for runoff and 58.1% for chemical denudation. Inherent variability in short term records gives rise to two major difficulties: first, in obtaining representative rates of process operation for a given environment in order to examine regional and climatic controls and second, in obtaining realistic magnitude and frequency data. Further complications inevitably arise in selecting the most suitable method for sediment yield estimation (cf. Walling, 1978; Ferguson, 1986, 1987).

In addition to the problem of obtaining data at the relevant timescale, contemporary reviews of erosion and sediment yield research highlight the need to link the fluvial environment to the hillslope and valley floor (cf. Hadley *et al.*, 1985). This linkage is likely to prove difficult from studies of fluvial processes alone yet it has been stressed by Dunne (1984) that attempts should be made to identify sediment sources and sinks within studies of fluvial sediment transport.

LAKE AND RESERVOIR SEDIMENTATION

Reservoir resurvey methods have been used for many years to estimate the expected life of the reservoir and catchment sediment yields (cf. Brown, 1944). A comprehensive range of survey and resurvey techniques have been developed to calculate the sediment volumes deposited (Bruk, 1985; Pemberton & Blanton, 1980; Rausch & Heinemann, 1984; Vanoni, 1975). Many studies based on resurvey and/or remote sensing methods report average yields for the entire life of the reservoir (e.g. McManus & Duck, 1985) but developments in the ^{137}Cs dating of reservoir sediments reported by Ritchie *et al.* (1973) have enabled some estimate to be made of variations in deposition rate through time (cf. Batten & Hindall, 1980).

Despite these and other technical developments, and the strengthening of conceptual links between lakes and their contributing catchments (Oldfield, 1977), a number of important practical difficulties still remain in utilizing the bottom sediments of lakes and reservoirs for reconstructing sediment yield histories. These include the identification and/or quantification of sediment source, trap efficiency, resuspension processes, sediment density changes, autochthonous and allochthonous contributions to the sediment and the significance of authigenic and mixing processes. Furthermore, refinements in sediment retrieval techniques and the methods employed to derive an absolute chronology must be seen as essential to the accurate quantification of sediment yield histories. These problems are briefly outlined below.

Sources of sediment

Hakanson & Jansson (1983) identify four major factors which control sedimentation in lakes. First, a depositional factor which expresses the capacity of a lake to act as a sediment trap (all other things being equal, larger lakes tend to be more efficient sediment traps). Secondly, the lake or reservoir will have its own internal productivity which will contribute towards the accumulation of sediment at the lake bed. Thirdly, pretrapping of sediment in upstream lakes and reservoirs will limit sediment supply. Fourthly, the natural load factor derived from allochthonous inputs (direct drainage basin derived inputs or atmospheric contributions to the lake surface). Any attempt to utilize the lake sediment record must be capable of distinguishing the respective contribution of autochthonous and allochthonous material and the relative contribution derived from the drainage basin and the atmosphere.

Trap efficiency

Sediment trap efficiency has attracted considerable attention from hydrologists. Graf (1983), for example, has shown that the general pattern of sedimentation is a function of changing hydraulic conditions, with the relatively high velocity turbulent inflow being transferred to slow flowing water within the lake or reservoir. Coarser particles, including the bedload, are usually deposited as a delta whilst the lighter particles, especially fine silts and clays, are distributed further into the water body. The exact distribution of the sediment will depend on factors such as the relative densities of the inflowing river and lake waters and the position of the thermocline (or pycnocline) if one exists (Fig. 1(a)). Furthermore, the chemical properties of the water in which settling takes place may enhance or inhibit settling through the impact of the sodium adsorption ratio on flocculation (e.g. Trujillo, 1982).

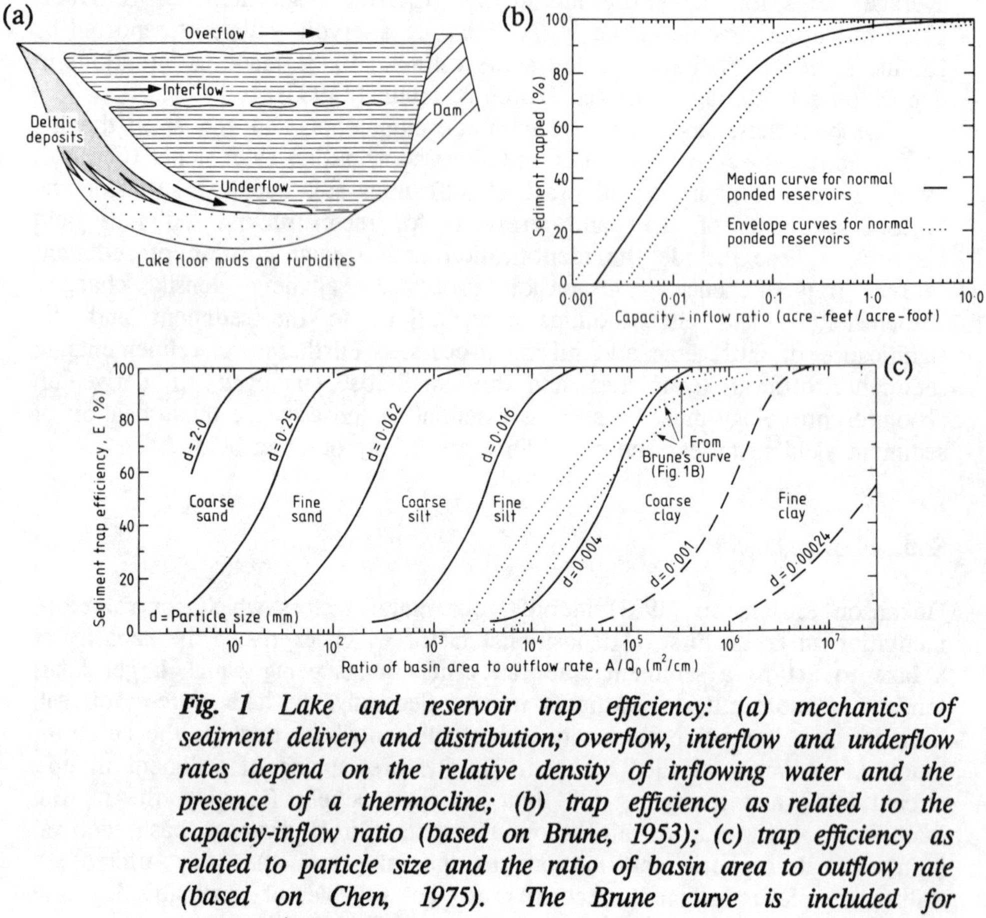

Fig. 1 Lake and reservoir trap efficiency: (a) mechanics of sediment delivery and distribution; overflow, interflow and underflow rates depend on the relative density of inflowing water and the presence of a thermocline; (b) trap efficiency as related to the capacity-inflow ratio (based on Brune, 1953); (c) trap efficiency as related to particle size and the ratio of basin area to outflow rate (based on Chen, 1975). The Brune curve is included for comparative purposes.

The efficiency with which reservoirs trap sediment can be predicted in a number of ways. Brown (1944) suggests that trap efficiency (*Te*) can be determined from:

$$Te(\%) = 100 \times 1 - \frac{1}{1 + 0.1C/DA}$$

where C = reservoir capacity; and
DA = basin area.

The drainage area parameter, however, seems to be a relatively poor substitute for inflow volume, and Brune (1953) developed the use of the capacity inflow ratio in preference to the capacity drainage area ratio (Fig. 1(b)). In his survey of 44 reservoirs, Brune found a range of trap efficiencies. Reservoirs with low capacity inflow ratios may fill and scour depending on the pervading streamflow conditions, whereas high retention capacities are to be found in reservoirs with high capacity inflow ratios, where continuous sedimentation is experienced and clear water is released downstream. In small basins, the trap efficiency curve developed by Heinemann (1981) may be more appropriate or, where a particle size differentiation is seen to be important, the capacity inflow/particle size relationship may be relevant (see Fig. 1(c); Chen, 1975; Heinemann, 1984; and Rausch & Heinemann, 1984). Trap efficiency, although a vital component for quantitatively estimating sediment yield, may vary in the same lake or reservoir depending on inflow conditions. For example, the trap efficiency of the Aswan high dam for a 15 year period between 1964 and 1979 varied from 84.8% to 99.9% (based on data in Shalash, 1982). For the purposes of sediment yield estimation, lakes and reservoirs with trap efficiencies approaching 100% will provide optimum sites for sediment yield studies.

Resuspension

Despite the relatively efficient trapping of sediments in some lakes and reservoirs, the resuspension, redeposition and sediment focussing process has a significant bearing on the methods which may be employed to estimate sediment yield. These problems have been investigated in recent limnological research, for example by Davis (1974) and Davis & Ford (1982) in Mirror Lake, New Hampshire and in Esthwaite Water in the UK (Hilton, *et al.*, 1986). Resuspension appears to be most relevant to the use of seston traps for estimating sediment accumulation rates in lakes and reservoirs (cf. Bloesch & Burns, 1980; Blomqvist & Hakanson, 1981), although it may also have implications for the preservation of the radioisotope record as discussed below. In a more general review of the problem, Hilton (1986) suggests that four processes dominate the resuspension and potential focussing of lake sediments. These include peripheral wave attack, random redistribution, intermittent complete mixing and slumping and sliding on slopes. Indeed, Davis *et al.* (1984) have argued that little information is as yet available on the understanding of the hydrodynamic and sedimentological processes which control sediment deposition in small lakes, although some attempts to model the process have been made on the basis of lake morphology (e.g. Lehman, 1975). An inability to predict the process of sediment focussing has implications for the design of sediment survey techniques, since basic morphometric properties cannot be used to predict the points of maximum, minimum and, more importantly, average sedimentation for a lake basin. As Dearing (1983) has shown in a small Scanian lake, the point of average sediment accumulation at the lake bed may vary over time depending on changes in exposure conditions or in response to local depositional processes.

Sediment density

In many cases, reservoir resurvey techniques have been used to estimate the volumetric accumulation of sediment for economic as well as geomorphological reasons (e.g. Stromquist, 1981; Bruk, 1985; McManus & Duck, 1985). Although the resurvey technique may be of value in assessing reservoir life, it is suboptimal for assessing sediment yield for a number of reasons. Firstly, sediment density may not be measured directly and may be assumed or estimated from one of the available empirically-derived formulae. Secondly, without sediment cores for analysis, the relative proportions of the autochthonous/allochthonous components cannot be estimated. Thirdly, the sediment yield estimate may span the entire life of the reservoir covering several periods of human impact or change. Fourthly, initial surveys following construction may be inadequate to provide a baseline against which to adequately assess subsequent deposition, since early adjustments might involve a greater proportion of bank derived sediments.

Of particular importance in sediment yield estimation is the change in density which can occur either during deposition or in post-depositional diagenesis. The density of sediment can be estimated from one of two major approaches. Reservoir engineers, for example, frequently consider the compaction of sediment in terms of the removal of pore water through time, assuming that the increase in compaction is time and/or particle size dependent. The methods most commonly used, according to Vanoni (1975), include that of Lane & Koelzer (1953) where the sediment density (p) after t years is given as:

$$p = p_1 + k \log t$$

where: p_1 is the sediment density after 1 year, k is a constant and t is time since deposition in years.

The detailed considerations given by Bolton (1986) to the above and other equations are too lengthy to be presented in detail here, but after reviewing a range of models based on inadequately formulated hydraulic principles, Bolton suggests that a mathematical formulation of the density profile, such as:

$$p = p_f - a\, e^{by}$$

where: p is a sediment density, a and b are constants which can be obtained from curve fitting techniques, p_f is a hypothetical maximum density value and y is depth; may be more appropriate.

The limitations of this approach are seen in hydraulic terms to relate to inadequate pore pressure dissipation during consolidation, although the problem may not manifest itself in slowly sedimenting basins where this process may equilibrate through time. However, the assumption that the density will tend towards a finite maximum is invalidated by the observations in a 500 000 year continuous sedimentary sequence in Lake Biwa, Japan (Yamomoto, 1984) where density increases with depth through over 200 m of sediment. Such records may form an important basis for the evaluation of a variety of models of sediment density.

An alternative approach to the density problem is presented by Hakanson & Jansson (1983). They assume that water content is the key component which they define as:

$$W = \frac{gws - gds}{gws}$$

where *gws* and *gds* are wet and oven dry masses respectively from a known sample volume.

Water content will decline with depth in the form of a negative exponential. More importantly, water content is also related to wet bulk density (*pw*), organic content and the density of organic material. Although humus may have a density close to water (1.3 to 1.5 g cm^{-3}), the density of minerogenic sediments may exceed 5.0 g cm^{-3}. Clearly, the mixture of sediments of various types affect the final density, not simply because of pore pressure and particle size differences, but also as a result of differential organic:inorganic loadings. Analyses of the organic matter content of lake sediments have shown that values may range from a few percent to over 30% by weight (Engstrom & Wright, 1984) and may reflect the organic productivity in the epilimnion of stratified lakes receiving high nutrient loadings. A recent investigation by Foster & Dearing (1987) has shown that in two morphologically and geologically similar reservoir basins in the Midlands of England, the average organic matter content of the sediments in the two basins ranged from 8 to 15% of dry sediment mass.

The accuracy of sediment yield estimates for whole lake basins may be inadequate, not only because of errors in estimating the vertical variations in density but also because spatial variations in density over the lake or reservoir bed will relate to inflow and secondary sorting processes as well as to local variations in erosion and deposition. For example, Smith *et al.* (1960) have shown that dry densities of surficial sediments in Lake Mead ranged from 0.53 to 0.91 t m^{-3} (excluding deltaic deposits). Such a range of densities would produce a significant variation in the estimated sediment mass and any consequent calculation of sediment yield.

The assumptions behind many of the hydraulic models reviewed by Bolton (1986) are also invalidated for a number of reasons, as illustrated by the variable nature of short core densities in a range of environments (Fig. 2). All of these profiles (except Lake Bussjo which was sampled with a "Russian Corer") were obtained from samples extruded from a Mackereth corer (Mackereth, 1969). Despite some of the problems in coring techniques outlined below, a number of important conclusions can be drawn with regard to these cores. For example, Fig. 2(a) is derived from a reservoir in Midland England and exhibits a relatively uniform increase in density with depth which appears to conform best to hydraulic models of compaction. The dramatic increase at 1 on the graph represents the corer penetrating the clay liner to the reservoir. Figure 2(b) refers to an upland lake in north Wales showing evidence of density increases (1 and 2) as a result of the inwash of mining

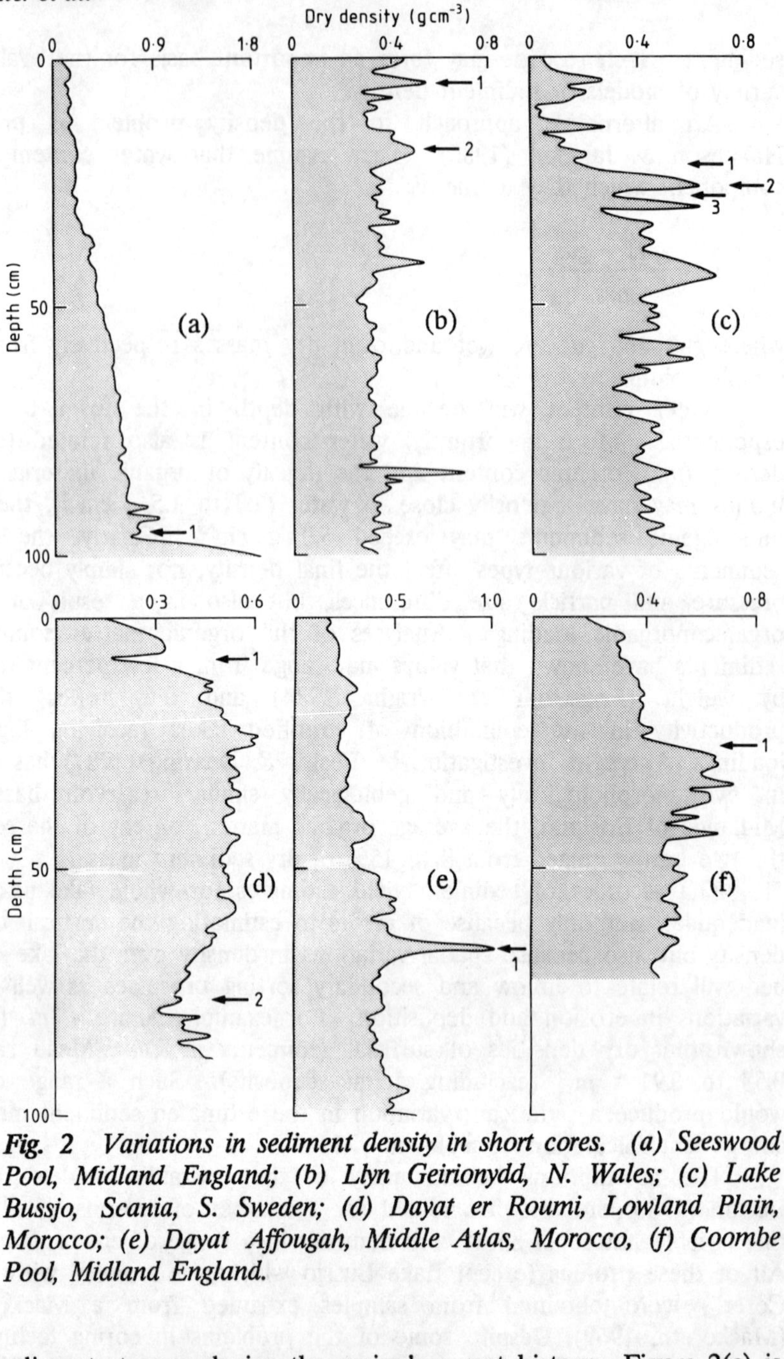

***Fig. 2** Variations in sediment density in short cores. (a) Seeswood Pool, Midland England; (b) Llyn Geirionydd, N. Wales; (c) Lake Bussjo, Scania, S. Sweden; (d) Dayat er Roumi, Lowland Plain, Morocco; (e) Dayat Affougah, Middle Atlas, Morocco, (f) Coombe Pool, Midland England.*

waste from adjacent streams during the region's recent history. Figure 2(c) is the density profile for a small kettle-hole lake in Scania, south Sweden, typified by alternating biogenic (3) and minerogenic (2) deposition producing a rhythmic density profile. The dramatic decrease in density at 1 on Fig. 2(c) is thought to represent a period of drainage during the 1950s and a reduction in effective sediment contributing area. Figure 2(d) illustrates the impact of a change of sediment source in a lake-catchment in Morocco, where the increase in density at around 70 cm depth is associated with the initiation of gully development (see Flower *et al.*, 1984; Flower *et al.*, 1989; Foster *et al.*, 1986a). Figures 2(e) and 2(f) reflect water level lowerings. Figure 2(e), which refers to a small lake in the Middle Atlas of North Africa, shows a peak (1) reflecting the presence of a high density shell layer thought to represent a period of lower water levels whilst Fig. 2(f) relates to a small reservoir in Midland England where an increase in density at *c.* 20 cm depth has been caused by drawdown for an extended period of around 4 years in the 1940s. Clearly, the range of factors responsible for controlling sediment density, which will also include particle size considerations, precludes the prediction of density with any degree of certainty. At the present time, there appears to be little substitute to the application of coring methods to quantify accurately spatial and post-depositional variations in sediment density. (These methods are discussed below).

Autochthonous and allochthonous sources

Lake sediments will comprise material derived from erosional processes within the upstream drainage basin, lake marginal erosion processes, atmospheric sources and biotic processes within the circulating water body which may selectively assimilate elements delivered to the lake in solution. Some attempt to distinguish these sources must be made where they are likely to contribute significantly to the accumulating sediment. Little has yet been done on the possible significance of lake marginal erosion processes, where imperceptible backwearing may make significant contributions to the accumulating sediment. This problem has been investigated theoretically by Dearing & Foster (1986) who produced the nomogram shown in Fig. 3 in an attempt to quantify the maximum acceptable bank erosion rate in a range of recent lake sedimentbased estimates of sediment yield. The factors controlling bank erosion rates will include wave height, height of erodible shoreline, fetch, local water level lowerings and the magnitude of currents capable of transporting eroded sediments to the central part of the lake or reservoir. Water level lowering is more problematic in reservoirs, and many studies have evidenced the secondary erosion/depositional sequences associated with the reworking of deltaic or other sediments (cf. Bruk, 1985; Szechowycz, 1973). Particularly problematic is the deliberate use of scour valves to remove reservoir bottom sediments. Such disturbances are likely to invalidate the lake sediment based method of sediment yield estimation. It is recommended that some attempt be made to monitor contemporary rates of bank retreat in order to quantify the magnitude of this problem in a variety of locations. Erosion pin studies are currently being carried out by the authors on a reservoir site in Midland England in order to quantify this problem.

The contribution derived from atmospheric inputs is usually assumed to be negligible. Records from Midland England have, however, shown that localized dust fallout could exceed 30 t km^{-2} $year^{-1}$ in recent times. Undoubtedly, these fallout records are unlikely to represent regional rates of

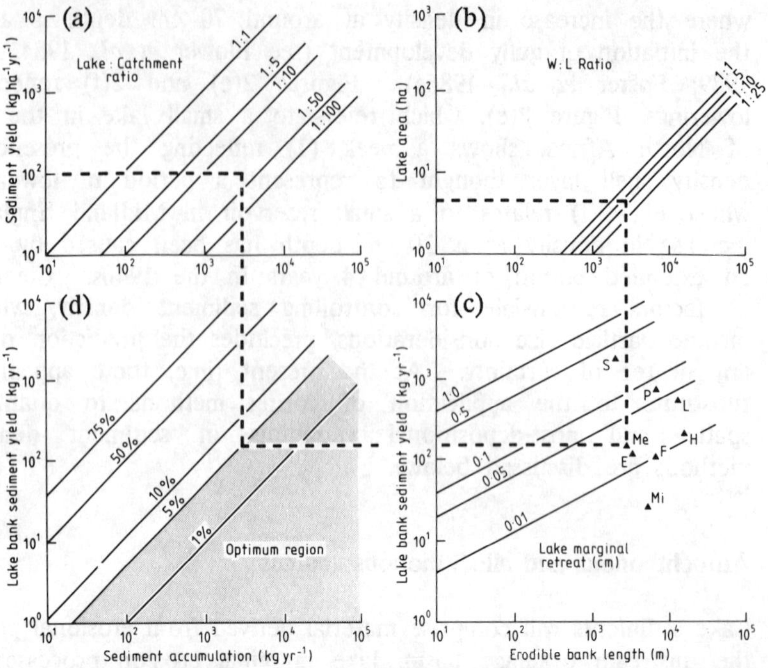

***Fig. 3** The impact of lake bank erosion on yield estimate: (a) relationship between lake:catchment area ratio and sediment yield to estimate dry sediment accumulation (kg yr^{-1}); (b) relationship between lake area and lake width to length (W:L) ratio to give estimate of erodible bank length; (c) relationship between erodible bank length, rate of retreat and sediment yield derived from bank erosion (assuming a bank height of 40 cm and a sediment density of 2.65 g cm^{-3}); (d) relationship between lake bank sediment yield, sediment accumulation and the % contribution to total yield by bank erosion.*

The nomogram requires two sets of input, a regional rate of sediment yield (Fig. 3(a)) and lake area and the W:L ratio (Fig. 3(b)). The example of its use, given for the dashed line of Merevale Lake, Warwickshire, shows a regional estimate of sediment yield of 100 kg ha^{-1} year^{-1}. In order for the lake marginal retreat to have less than a 5% contribution to sediment yield, a retreat of between 0.01 and 0.05 cm year^{-1} must be assumed. The following lakes are plotted: Merevale Lake (Me, Foster et al., 1985); Seeswood Pool (S, Foster et al., 1986b); Frains Lake (F, Davis, 1976); Loe Pool (L, O'Sullivan et al., 1982); Llyn Peris (P, Dearing et al., 1981); Havgardssjon (H, Oldfield et al., 1983); Mirror Lake (Mi, Likens & Davis, 1975); and Lake Egari (E, Oldfield et al., 1985) (based on a diagram in Dearing & Foster, 1986).

deposition, but attempts by Foster *et al.* (1985) to quantify this component in a slowly sedimenting Midland England reservoir suggests that the atmospheric input could contribute up to 9% of the gross sediment accumulation in recent years.

Estimation of the contribution made by internal productivity can be made by an analysis of the organic content of the accumulating sediment, which may be compared with the organic content of contemporary inflowing sediments. Hakanson & Jansson (1983) have empirically modelled the relationship between density and organic matter content, based on the weight loss recorded when sediments are ignited in a muffle furnace at low temperatures (540°C). Such analyses may only record a small component of the productivity controlled deposition, since diatom frustules composed of silica may constitute a large proportion of the sediment mass. Here, the contribution may be estimated by analysis of the accumulating sediments by an alkaline digestion procedure (cf. Engstrom & Wright, 1984 and Foster *et al.*, 1985, 1986b). Midland reservoirs in the UK may have as much as 30% of their sediment mass accounted for by organic matter and biogenic silica. Another particularly problematic area relates to the secondary precipitation of calcite in lake waters in lakes of high salinity or where soils are carbonate-rich. Extraction techniques seem unable to differentiate between autochthonous and allochthonous calcium carbonate and as far as is currently known, no lake sediment based estimate of sediment yield has yet attempted to deal directly with this problem. Several approaches may be adopted, such as to undertake comparative extractions in soils and lake sediments to assess the gross changes in calcium carbonate content, to undertake XRD analysis to quantitatively distinguish between calcite and dolomite in the sediments and sources, or to adjust for all $CaCO_3$ contents by expressing yields on a carbonate free basis. Other gypsiferous or sodium-rich deposits in arid and semiarid lakes may be more easily distinguished in the lake sediment record and their relative contribution adjusted accordingly. Other autochthonous components may be quantified by means of a variety of chemical methods discussed by Engstrom & Wright (1984) and Bengtsson & Enell (1986).

Authigenic and mixing processes

Of less significance for sediment accumulation studies, but of great relevance to the preservation of chemical and radiometric stratigraphies, is the variable nature of the combined effect of bioturbation and chemical diffusion. These processes are relevant here in that the separation of the sedimentary record into intervals approaching a decade in resolution is wholely dependent upon the mechanisms responsible for the delivery, adsorption and diagenesis of those isotopes which form the basis of radiometric chronologies. The two isotopes most commonly employed for short core studies are ^{137}Cs and ^{210}Pb. For timescales exceeding 100–200 years, varved, ^{14}C and/or palaeomagnetic chronologies may be more appropriate (see below). Few models as yet exist to quantify the significance of these components, but Fisher *et al.* (1980) in a series of laboratory experiments have shown how *Tubifex tubifex* incubated at different levels in the sediment could affect the movement and diffusion of ^{137}Cs in a laboratory tank experiment and Davis (1974) demonstrated that feeding depths of tubificids reached 15cm in the profundal sediments of Messalonskee Lake, Maine. More recently, Hakanson & Jansson (1983) have proposed a dynamic model of the process, which incorporates parameters such as sediment depth, rate of sedimentation, water content, bulk density, compaction, biotransport, substrate decomposition and bioturbation limit. Such a model may form the basis for assessing the importance of density variations as well as the bioturbation process on sediment yield estimation.

BOTTOM SEDIMENT SAMPLING

The selection of an appropriate sampling framework in order to provide accurate and precise estimates of sediment yields involves several problems. To date, most studies have performed multiple corings in order to account for the spatial variation in deposition rate. The coring densities associated with studies with which the authors have been involved range from 1 per 0.0035 ha in Llyn Goddionduon to 1 per 12.1 ha in Dayat er Roumi, Morocco. Some examples of coring densities are given in Table 1.

The sampling framework is usually arranged on a rectangular grid to

Table 1 Coring densities in bottom sediment studies

Site and Location	*Lake area (ha)*	*Coring density (ha per core)*
Goddionduon (1) North Wales	*6.2*	*0.0035 - 0.04*
Frains Lake (2) Michigan	*6.7*	*0.3*
Havgärdssjön (3) Scania, Sweden	*55*	*1.0*
Llyn Peris (4) North Wales	*20*	*1.3*
Merevale (5) Midland England	*6.5*	*0.08*
Seeswood (6) Midland England	*6.7*	*0.16*
Egari (7) New Guinea	*8.5*	*1.7*
Loe Pool (8) Southwest England	*44*	*1.0*
Geirionydd North Wales	*30*	*0.87*
Catherine Northern Ireland	*36*	*1.5*
Llangorse Mid Wales	*150*	*3.0*
Roumi(9) Morocco	*85*	*12.1*
Affougah (9) Morocco	*6*	*1.2*
Azigza (9) Morocco	*37*	*7.4*
Bussjo Scania, Sweden	*0.84*	*0.12*

1 = Bloemendal et al. 1979; 2 = Davis 1976; 3 = Dearing 1986; 4 = Dearing et al. 1981; 5 = Foster et al. 1985; 6 = Foster et al. 1986; 7 = Oldfield et al. 1985; 8 = O'Sullivan et al. 1982; 9 = Flower et al. (in prep).

improve the speed of sample point location and to ensure that a representative range of sedimentary environments is sampled. With small lakes and reservoirs, such as those listed in Table 1 (largest 150 ha) it is often appropriate to locate sampling stations within the lake either by stakes pushed into the bottom sediments or by buoys anchored at sampling points. Coring positions are identified at the intersection of marked shore stations located by back bearings from the boat to the shore. Where practicable, ropes can be stretched across small lakes and reservoirs between shore stations laid out by local ground survey. Exact coring positions may be located on each transect by measuring distance across the lake. Further refinements may be introduced with the use of automated techniques for sample site positioning, especially on large water bodies (e.g. Lambert, 1982; Battarbee *et al.*, 1983; Bruk, 1985). On lakes which are seasonally frozen, cores may be retrieved through the ice and traditional land survey techniques may be used to locate sampling positions.

Sediment retrieval

Numerous techniques have been developed in order to retrieve bottom sediments from lakes and reservoirs. The exact method selected will depend on the following criteria:

(a) depth of water;
(b) whether the water surface is frozen;
(c) stability of the coring platform;
(d) thickness of the sediment to be cored;
(e) cohesive properties of the sediment;
(f) whether an undisturbed surface is to be retrieved;
(g) whether undisturbed samples are to be subsampled in the field;
(h) mass of sediment required for subsequent analysis; and
(i) whether sediments are laminated.

Various sampling methods have been reviewed, for example, by Wright (1980) and Aaby & Digerfeldt (1986). Figure 4 illustrates some of the basic principles of corer operation. For example, Fig. 4(a), the "Russian" corer, is a chamber corer which samples undisturbed material. However, it can only be operated from a frozen surface or a stable raft or platform and, with deep water (>3–4 m), it requires guide tubes to avoid bending the rods. It is unsuitable for low density surface sediments, but in higher density materials it may retrieve a sufficiently undisturbed sample for density and palaeomagnetic determinations. In deeper water where rod operation is impractical, a line operated piston type sampler may be more appropriate (Fig. 4(b)). The fixed line holds a Kullenberg (1947) seal above the sediment surface and the piston is driven past the seal into the sediment with a line operated weight. The partial suction created by the seal prevents the sample falling out of the piston as it is raised to the surface. The "frozen finger" type sampler (Fig. 4(c)) is a gravity operated device which is filled with dry ice and alcohol. The sediment freezes to the outer surface of the corer which is retrieved with the hand line. Although the sampler may be unsuitable for density analysis, the

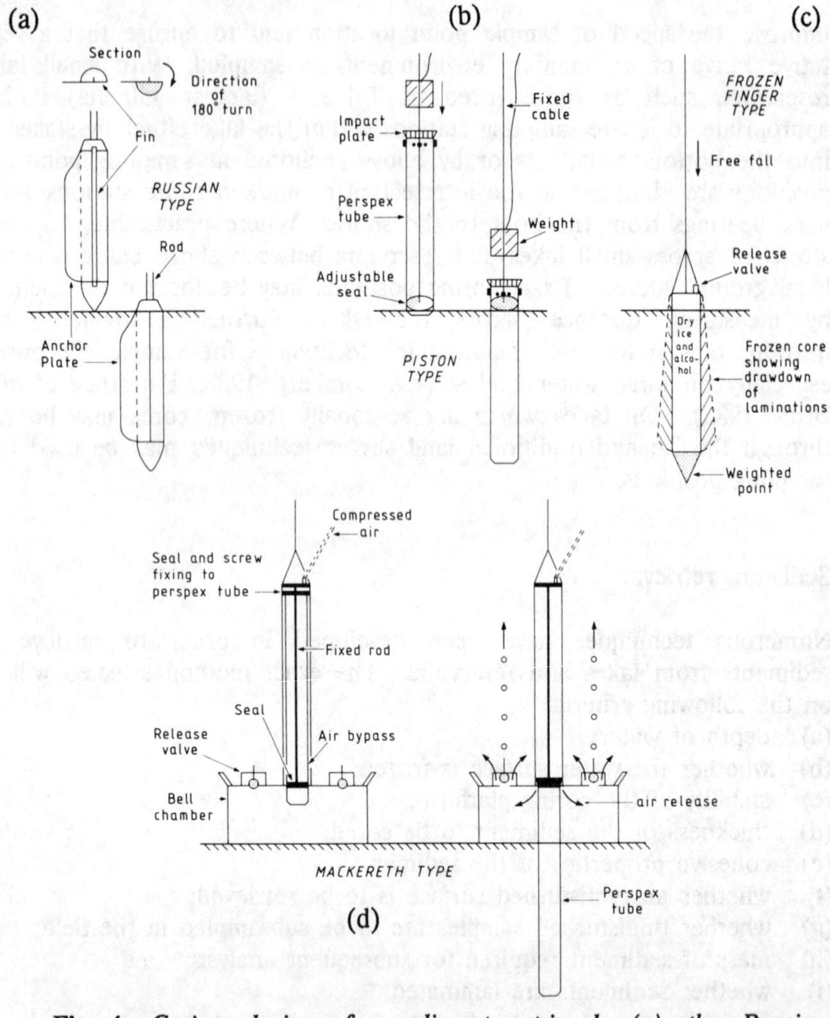

***Fig. 4** Coring devices for sediment retrieval: (a) the Russian type; (b) the piston type; (c) the frozen finger type; and (d) the Mackereth (pneumatic) type.*

technique is particularly suitable for the study of laminated sediments (cf. O'Sullivan *et al.*, 1982). One of the most important developments has been the availability of compressed air driven samplers (cf. Mackereth, 1969) which are capable of retrieving an undisturbed sediment surface as well as continuous cores of up to 6 m of undisturbed material from a variety of depths and, theoretically, in any depth of water exceeding the length of the corer. The principle of operation of the "mini corer" is shown in Fig. 4(d). A bell chamber provides a stable platform for the coring operation and the piston is pushed past a seal with compressed air pumped into the corer from the boat. The corer penetrates the sediment and air bubbles released through the release valves indicate completion of coring at the water surface. The corer is retrieved by hand. For 3 and 6 m versions of this sampler, the system is modified to "suck" the corer chamber into the sediment in order to improve stability and, once coring is complete, this chamber fills with compressed air to provide lift. Both corers can be operated from small boats. One of the greatest limitations of the piston samplers, including those driven by compressed air, is that they are unreliable in sediments with low cohesion and high water contents. Estimates of density are subject to error in the Mackereth and similar types of piston corer because some compaction, foreshortening and even sediment loss may occur during sampling (Blomqvist, 1985) and vertical extrusion of cores with a piston may also lead to some compaction.

Core correlation and dating

Since sedimentation rates vary across the lake or reservoir bed, an important aspect of sediment yield reconstruction is the identification of time synchronous layers within the sediments, in order to calculate sediment volume. The solution to this problem demands some means of both core correlation and dating. Although dating methods may be applied to all cores, the prohibitive cost and time consuming nature of the analysis will frequently preclude more than one or two dated cores per lake. The techniques of core correlation have recently been reviewed by Dearing (1986) who suggested that correlation of synchronous levels between sediment cores demands that the property used for correlation should be areally continuous and synchronous. Various methods of core correlation have been used in lake and reservoir studies and these are summarized in Table 2. One of the more important developments in core correlation in recent years has been the measurement of the magnetic properties of lake sediments (see Thompson & Oldfield, 1986). These properties were shown to be particularly useful because of the speed of measurement in the first study of this type (Bloemendal *et al.*, 1979) and a range of magnetic properties have been used in many subsequent lake sediment based studies of sediment yield (e.g. Dearing *et al.*, 1981; Dearing, 1986; Foster *et al.*, 1985, 1986b). More recent investigations by Hilton & Lishman (1985) have, however, shown that some properties, such

***Table 2** Some methods of core correlation*

*1 **Visible Stratigraphy***	*e.g. annually laminated sediments (Simola et al., 1981); changes in texture and colour (e.g. Digerfeldt, 1976); tephra layers (Thompson et al., 1986).*
*2 **Palaeoecological Stratigraphy***	*e.g. Pollen horizons (Digerfeldt, 1974; Davis, 1976); diatom assemblages (e.g. O'Sullivan et al., 1973)*
*3 **Chemical Stratigraphies***	*e.g. loss on ignition (Tolonen et al., 1975; Davis, 1976).*
*4 **Radioisotope Record***	*e.g. ^{137}Cs record in surface sediments (Pennington et al., 1976).*
*5 **Magnetic Correlations***	*based on natural remanent and induced magnetic properties (see Thompson & Oldfield, 1986 and text for detail).*

as magnetic susceptibility may be controlled by a diagenetic process under differing redox conditions and may not be controlled by allogenic inputs to the lake basin. Some care should be exercised in selecting the most appropriate basis for magnetic or other core correlation methods, in order to adhere to the principles of areal continuity and areal synchroneity.

Provision of an accurate chronology is undoubtedly the most important aspect of sediment yield reconstruction and a variety of techniques have now been developed. These techniques may be based firstly, on the preservation of palaeomagnetic properties of inclination, declination and intensity, calibrated on a regional basis with radioisotope ages; secondly, on the production of natural isotopes in the environment which decay at a known rate relative to a stable form, such as ^{14}C and ^{210}Pb; thirdly, on the presence of an isotope such as ^{137}Cs which was introduced into the environment from atmospheric weapons testing; and, fourthly, on the existence of rhythmic or annual laminations or varves (cf. O'Sullivan, 1983). Space does not permit a full discussion of the technical problems involved and a number of recent reviews have dealt with palaeomagnetic methods (Thompson & Oldfield, 1986; Thompson, 1986) and radiometric dating including weapons testing isotopes (Cambray *et al.*, 1982; Lowe & Walker, 1984; Oldfield & Appleby, 1984; Olsson, 1986).

For reservoirs constructed over the last 200 years or less, a combination of ^{210}Pb and ^{137}Cs analyses seem to be most appropriate in providing resolution of a decade or less over the appropriate time period. Given the half life of ^{210}Pb of 22.26 ± 0.22 years, this radionuclide is particularly suited to this timescale and can potentially give accurate age determinations for up to 150 years. Recent investigations by Flower *et al.* (1989) in North Africa have, however, experienced some difficulty in obtaining a reliable chronology older than 30 years BP from this isotope in a lake basin which is accumulating at a particularly rapid rate.

One of the most important problems recently identified is the potential unreliability of the ^{137}Cs record and its apparent dependence not only on the bioturbation and mixing processes outlined above, but also on the potential downward molecular diffusion and adsorption of this ion. This problem was highlighted by Davis *et al.* (1984) in a number of Scandinavian and New England lake sediment cores which were dated by pollen marker horizons as well as by the ^{210}Pb method. The latter appears to be little affected by downward diffusion through the sediment column.

Some examples of the application of ^{137}Cs and ^{210}Pb for dating cores are given in Fig. 5 for two Midland England reservoirs. The degree of coincidence between the atmospheric and lake sediment records in Figs 5(a), 5(b) and 5(c) is variable and the correlation of these records with the ^{210}Pb record shown in Figs 5(d) and 5(e) depends on the model used to calculate the depth:age relationship in the sediment. At least two dated cores seem to be necessary to overcome the problems of sediment focussing discussed above and other independent means of core correlation between the two dated cores should be used in order to "fine tune" the depth:age curve in different parts of the basin (cf. Oldfield & Appleby, 1984; Fig. 5(e)).

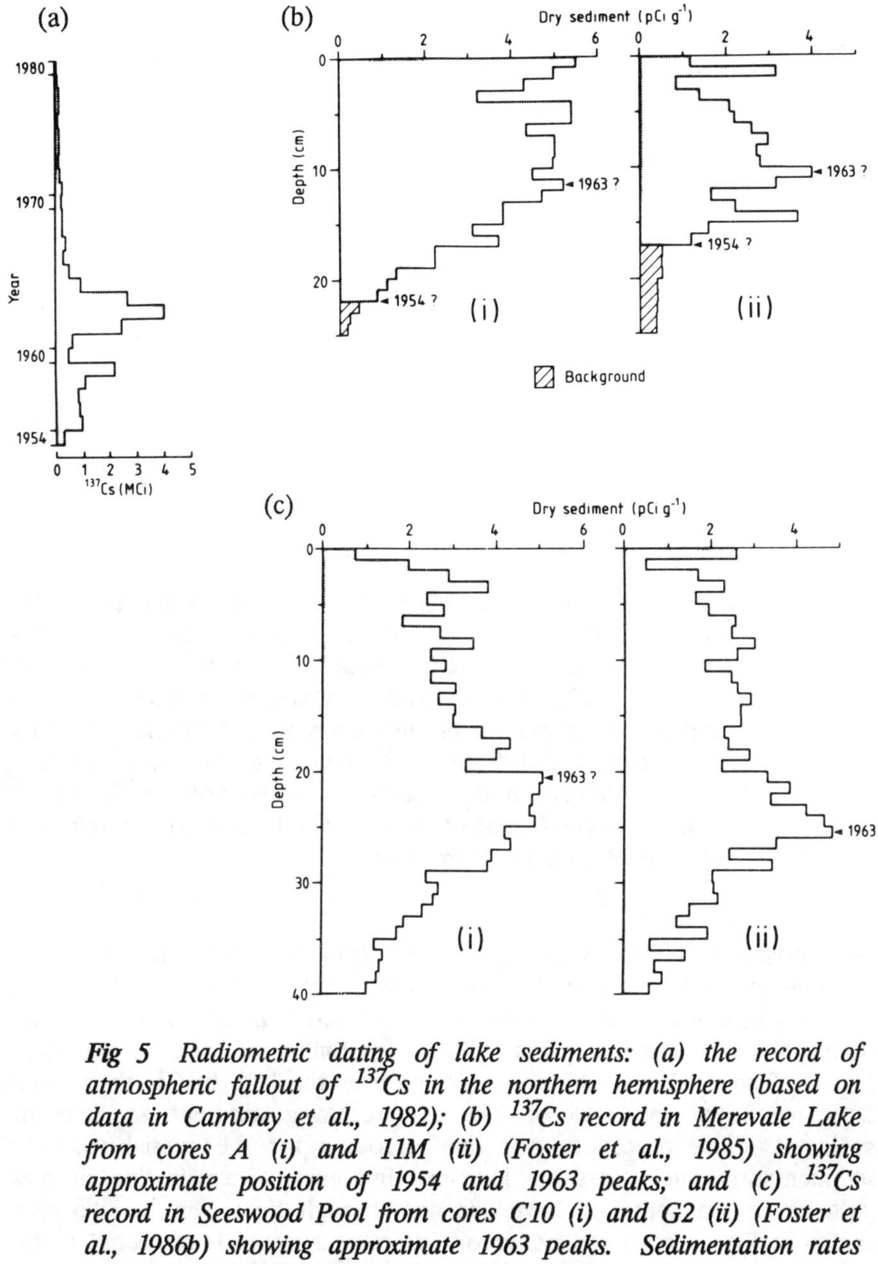

***Fig 5** Radiometric dating of lake sediments: (a) the record of atmospheric fallout of ^{137}Cs in the northern hemisphere (based on data in Cambray et al., 1982); (b) ^{137}Cs record in Merevale Lake from cores A (i) and 11M (ii) (Foster et al., 1985) showing approximate position of 1954 and 1963 peaks; and (c) ^{137}Cs record in Seeswood Pool from cores C10 (i) and G2 (ii) (Foster et al., 1986b) showing approximate 1963 peaks. Sedimentation rates in Seeswood pool are approximately twice those of Merevale in the upper part of these cores.*

SEDIMENT YIELD CALCULATION AND INTERPRETATION

The total influx of sediment to a lake or reservoir for each synchronous and dated horizon can be obtained by multiplying the associated mean wet sediment volume of each core by the percentage weight loss measured by

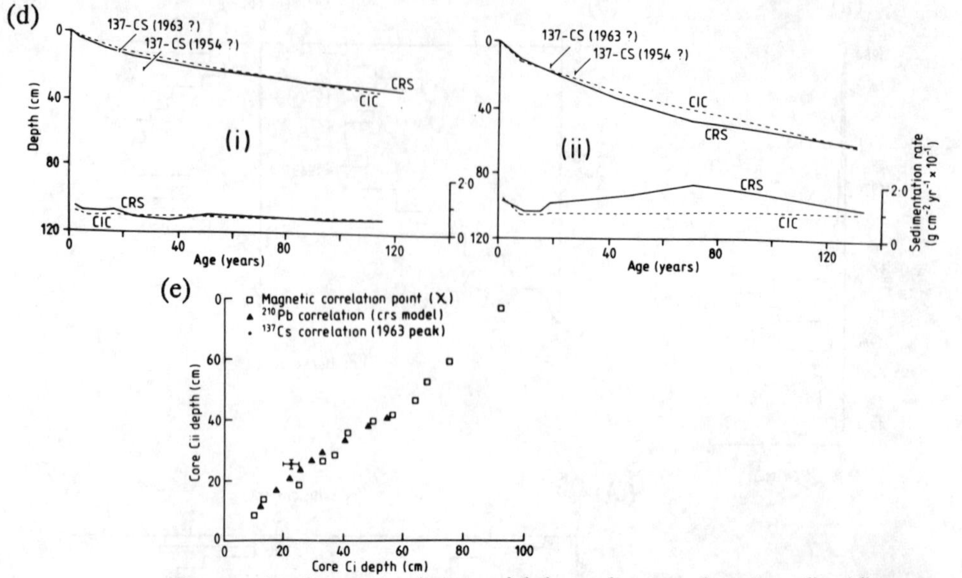

Fig. 5 Radiometric dating of lake sediments (continued): (d) comparison of the "Constant Rate of Supply" (CRS) and "Constant Initial Concentrations" (CIC) models for ^{210}Pb dating (see Oldfield & Appleby, 1984 for discussion); depth-age chronologies and sedimentation rates are for the two cores from Merevale lake shown in Fig. 5 (b)(i) and 5 (b)(ii), ^{137}Cs dates are also plotted for comparison; (e) comparison of radiometric chronologies (^{137}Cs and ^{210}Pb) for the Seeswood Pool cores (Fig. 5 (c)(i and ii)) as related to an independent magnetic correlation.

oven drying at 110°C, in order to obtain total dry sediment mass. A more accurate method is to use the dry sediment density data for all cores within each synchronous zone. A map of the dry sedimentation rate for each area of the basin can then be produced for each time zone, such as that shown in Fig. 6. The total mass of dry sediment over the lake bed is then calculated after measuring the area of lake bed receiving different amounts of dry sediment. This procedure not only accounts for the spatial variability in sedimentation rate across the lake bed but enables examination of how the patterns of sedimentation have changed through time. In the two examples shown in Fig. 6, early sedimentation in both reservoirs is restricted to the old river channels and valley floors. As time proceeds and these zones fill with sediment, the area receiving material expands towards the more marginal zones where later accumulation occurs.

As described above, the total mass of accumulating sediment is only partly a function of processes operating in the drainage basin and some attempt should be made to account for losses due to changing trap efficiency, and increases in mass caused by atmospheric input, lake marginal erosion and autochthonous and diagenetic contributions. These adjustments frequently demand chemical and other determinations on the retrieved sediment.

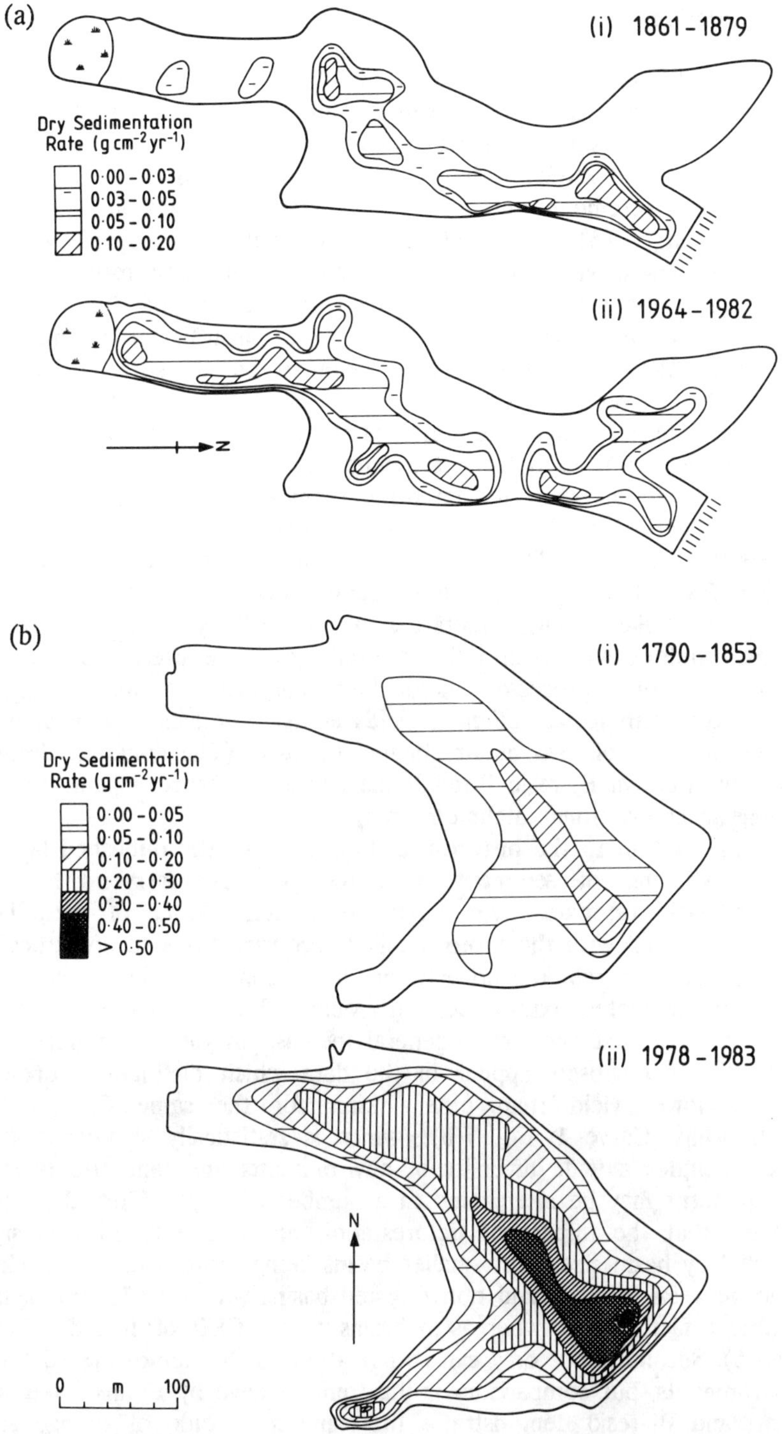

***Fig. 6** Change in sedimentation patterns in reservoirs: (a) Merevale lake; (b) Seeswood Pool, showing the expansion of sedimenting area as the valley floor infills in both reservoirs.*

Interpretation

Having subtracted the mass of sediment which is non-denudational, the remaining fraction should represent a value close to that calculated from river based sediment yields. Some attempt has been made to evaluate the correspondence between sediment yields calculated from river discharge and turbidity measurements and from reservoir sedimentation in Midland England (Foster *et al.*, 1985). This study has shown that within the likely variability in annual sediment yields demonstrated by the turbidity record, the adjusted lake sediment-based estimate of yield produces comparable results for the most recent time period. To date, however, insufficient emphasis has been placed on the comparison of lake sediment based estimates with estimates derived from river based studies.

To date, less than 10 continuous reconstructions of sediment yield, based on multiple coring of bottom sediments, have been published. These records cover a range of environments from Tropical lakes in highland Papua New Guinea (Oldfield *et al.*, 1985) to lowland lakes in seasonally cold environments of Southern Sweden (Dearing, 1986). Some of these records are reproduced in Fig. 7(a). Two types of lake are represented in this diagram. First, Frains Lake and Lakes Egari, Havgärdssjön and Bjäresjö are lakes receiving no channel inflow, where all inflowing sediments are presumably derived from surface erosional processes and/or lake marginal erosion. In contrast, the three lakes with lower sediment yields all have at least one and, in the case of Seeswood Pool, two major channel inflows. (The apparent decline in the sediment yield in Bjäresjö through the period of record appears to relate to a change in contributing catchment area).

In addition to the historical picture given by the published lake sediment based estimates of sediment yield, resurveys of reservoirs have provided average yield estimates over varying timescales. Some of these latter data have been used with the more detailed reconstructions to construct Fig. 2(b), which shows the relationship between the catchment to lake area ratio (CLR) and the estimated basin sediment yield. Two trends emerge from this relationship. First there is a general increase in yield with decreasing CLR. Secondly, two subsets appear in the data which distinguish forested basins with a lower yield than other basins for the same CLR. The upper relationship (Curve 1: Fig. 7(b)), which is statistically significant at the .001 level, includes a wide diversity of environments and land use types and the two patterns may be interpreted in a number of ways. First, the relationships suggest that the impact of deforestation on sediment yields is significantly affected by basin size, with smaller basins being more sensitive to change (the ratio between forested and non forested basins with a CLR approaching 1.0 is around a factor of 10 whereas in basins with a CLR of 100, the factor is less than 5). Secondly, not only can this relationship be demonstrated for different environments, but comparison of the trends shown by Frains Lake, and Lakes Egari and Bjäresjö demonstrates the same magnitude of change appropriate to their CLRs. Furthermore, comparison of Merevale Lake and Seeswood Pool, which are both in Midland England and have almost identical CLRs but contrasting land uses, indicates a ratio of forested to deforested basins in

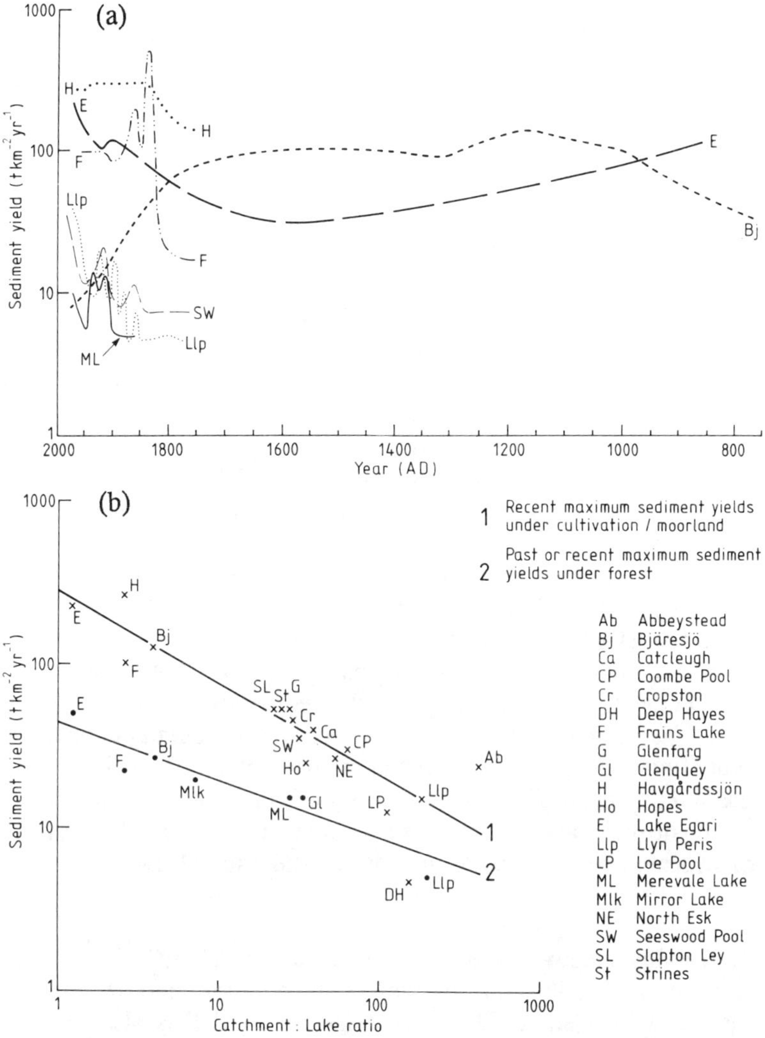

Fig. 7 ***Sediment yield estimates from lake and reservoir based studies: (a) some long term trends in sediment yield from lake sediment based studies; (b) sediment yields as related to the catchment to lake ratio (CLR). These include both lake sediment based and reservoir resurvey data, making allowance for water content and density.***

Data from the following published sources: Crick (1985), Davis (1976), Dearing (1986), Dearing et al. (1981), Cummins & Potter (1972), Foster et al. (1985), Foster et al. (1986), Hall (1967), Ledger et al. (1974), Likens & Davis (1975), McManus & Duck (1985), Oldfield et al. (1985), O'Sullivan et al. (1982), Rodda et al. (1976) and Young (1958).

accordance with the general relationships obtained. The CLR parameter is dominated by the influence of catchment size and it is suggested that it may closely relate to the sediment delivery ratio in fluvial studies (cf. Walling, 1983).

An ability to reconstruct patterns of sediment yield for a single environment is valuable for a number of reasons. Firstly, it is possible to exert some experimental control on the influence of catchment size on the computed result. (A paired lake catchment based study should account for the changing sensitivity of the environment at different CLRs). Secondly, careful selection of the basin enables testing of various models of the fluvial environment and general models of landscape sensitivity to change by using historical data to conduct experiments on our behalf (cf. Deevey, 1969). The contemporary analogue model produced by Wolman (1967) is frequently reproduced in fluvial texts, yet the quantitative reconstruction of sediment yields following deforestation presented by Davis (1976) rarely appears in the hydrological literature. This latter study evaluates not only the equilibrium conditions under forest and clearance but also quantifies the response to and recovery from a period of change. Conceptually, the latter is to be preferred and one might argue that the Wolman model should be modified in the light of these data. More recently, the lake sediment based record of sediment yield has been used to evaluate the important controls on sediment production in contrasting environments through an analysis of the relationship between historical rainfall records and sediment yield for the last two centuries (Dearing & Foster, 1987) and in association with tephra layers, the technique has been used by Thompson *et al.*, (1986) to calculate changes in the sediment input to Icelandic lake sediments.

The lake sediment based record of sediment yield is undoubtedly suboptimal for a detailed evaluation of contemporary process dynamics, but it has already been shown to add a significant dimension to the interpretation of sediment yield and sediment source data at a timescale relevant to the testing of hydrological and fluvial models of landscape change.

Acknowledgements We are indebted to NERC and NAB for funding research into lake and reservoir sedimentation and to Shirley Addleton for the usual high standard of the diagrams. IDLF is also grateful to the Royal Society, the Workshop Organising Committee and Coventry Polytechnic for providing funds to attend the workshop and present the paper.

REFERENCES

Aaby, B. & Digerfeldt, G. (1986) Sampling techniques for lakes and bogs. In: *Handbook of Holocene Palaeoecology and Palaeohydrology* (ed. B. Berglund), Wiley, Chichester.

Battarbee, R. W., Titcombe, D., Donnelly, K. & Anderson, J. (1983) An automated technique for the accurate positioning of sediment core sites and the bathymetric mapping of lake basins. *Hydrobiologia* **103**, 71-74.

Batten, W. G. & Hindall, S. M. (1980) Sediment deposition in the White River Reservoir, Northwestern Wisconsin. *USGS Wat. Supply Pap.* 2069.

Bengtsson, L. & Enell, M. (1986) Chemical analysis. In: *Handbook of Holocene Palaeoecology and Palaeohydrology* (ed. B. Berglund), Wiley, Chichester, UK.

Bloemendal, J., Oldfield, F. & Thompson, R. (1979) Magnetic measurements used to assess sediment influx at Llyn Goddionduon. *Nature* **280**, 50-53.

Bloesch, J. & Burns, N. M. (1980) A critical review of sediment trap technique. *Schweiz. Z. Hydrol.* **42**, 15-55.

Blomqvist, S. (1985) Reliability of core sampling of soft bottom sediment - an in situ study.

Sedimentology **32**, 605-612.
Blomqvist, S. & Hakanson, L. (1981) A review on sediment traps in aquatic environments. *Arch. Hydrobiol.* **91**, 101-132.
Bolton, P. (1986) *Estimating Sediment Densities in Reservoirs.* Tech Note OD/TN 17 Hydraulics Research, Wallingford, UK.
Brown, C. B. (1944) Sedimentation in reservoirs. *Trans. ASCE* **109**, 1085.
Bruk, S. (1985) (Rapporteur) *Methods of computing sedimentation in lakes and reservoirs.* Report of IHP-II Project A.2.6., UNESCO, Paris.
Brune, G. M. (1953) Trap efficiency of reservoirs. *Trans. AGU* **34**, 407-418.
Cambray, R. S., Playford, K. & Lewis, G. N. J. (1982) *Radioactive Fallout in Air and Rain: Results to End of 1981.* UK Atomic Energy Authority Harwell, AERE - R 10485.
Chen, C. N. (1975) Design of sediment retention basins. *Proc. National Symp. on Urban Hydrology and Sediment Control.* Univ. Kentucky, Lexington, USA. 58-68.
Crick, M. I. (1985) Investigations into the relationship between sediment accumulation in the lower Ley, Slapton and spatial patterns of erosion within its catchment using magnetic measurements. Unpublished BSc Dissertation, Dept of Environmental Science, Plymouth Polytechnic, UK
Cummins, W. R. & Potter, H. A. (1972) Rate of erosion in the catchment area of Cropston reservoir, Charnwood Forest, Leicestershire. *Mercian Geol.* **6**, 149-157.
Davis, R. B. (1974) Stratigraphic effects of tubificids in profundal lake sediments. *Limnol. Oceanogr.* **19**, 466-488.
Davis, R. B., Hess, T., Norton, S. A., Hanson, Hoagland, K. D. & Anderson, D. (1984) Cs-137 and Pb-210 dating of sediments from soft-water lakes in New England (USA) and Scandinavia, a failure of Cs-137 dating. *Chem. Geol.* **44**, 151-185.
Davis, M. B. (1976) Erosion rates and land use history in Southern Michigan. *Environ. Conserv.* **3**, 139-148.
Davis, M. B. & Ford, M. S. (1982) Sediment focusing in Mirror Lake, New Hampshire. *Limnol. Oceanogr.* **27**, 137-150.
Davis,, M. B., Moeller, R. E. & Ford, J. (1984) Sediment focusing and pollen influx. In: *Lake Sediments and Environmental History* (ed. E. Y. Hawarth & J. W. G. Lund) Leicester Univ. Press, Leicester, UK.
Dearing, J. A. (1983) Changing patterns of sediment accumulation in a small lake in Scania, southern Sweden. *Hydrobiologia* **103**, 59-64.
Dearing, J. A. (1986) Core correlation and total sediment influx. In: *Handbook of Holocene Palaeoecology and Palaeohydrology.* (ed. B. Berglund), Wiley, Chichester, UK.
Dearing, J. A., Elner, J. K. & Happey-Wood, C. M. (1981) Recent sediment influx and erosional processes in a Welsh upland lake-catchment based on magnetic susceptibility measurements. *Quatern. Res.* **16**, 356-372.
Dearing, J. A. & Foster, I. D. L. (1986) Lake sediments and palaeohydrological studies. In: *Handbook of Holocene Palaeoecology and Palaeohydrology,* (ed. B. Berglund), Wiley, Chichester, UK.
Dearing, J. A. & Foster, I. D. L. (1987) Limnic sediments used to reconstruct sediment yields and sources in the English Midlands since 1765. In: *Geomorphology '86* (ed. V. Gardiner), Wiley, Chichester, UK.
Deevey, E. (1969) Coaxing history to conduct experiments. *Bioscience* **9**, 40-43.
Digerfeldt, G. (1974) The post-glacial development of the Ranviken Bay in Lake Immeln. I: The history of the regional vegetation, and II: The water level changes. *Geo. Foren. i Stock. Forhand.* **96**, 3-32.
Digerfeldt, G. (1976) A pre-boreal water level change in Lake Lyngsjo, central Holland. *Geol. Foren. i Stock. Forhand* **98**, 329-336.
Dunne, T. (1984) The prediction of erosion in forests. Symp. Keynote Address. In: *Symp. on Effects of Forest Land Use on Erosion and Slope Stability* (ed C. L. O'Loughlin & A. J. Pearce). Environment and Policy Institute. East-West Center, University of Hawaii, Honolulu, Hawaii.
Engstrom, D. R. & Wright, H. E. (1984) Chemical stratigraphy of lake sediments as a record of environmental change. In: *Lake Sediments and Environmental History.* (ed. E. Y. Hawarth & J. W. G. Lund) Leicester Univ. Press. Leicester, UK.
Ferguson, R. I. (1986) River loads underestimated by rating curves. *Wat. Resour. Res.* **22**, 75-76.
Ferguson, R. I. (1987) Accuracy and precision of methods for estimating river loads. *Earth Surf. Proc. Landforms* **12**, 95-104.
Fisher, J. B., Lick, W. J., McCall, P. L. & Robbins, J. A. (1980) Vertical mixing of lake sediments by tubificid oligochaetes. *J. Geophys. Res.* **85**, 3997-4006.
Flower, R. J., Dearing, J. A. & Nawas, R. (1984) Sediment supply and accumulation in a small Moroccan lake: an historical perspective. *Hydrobiologia* **112**, 81-92.
Flower, R. J., Stevenson, A. C., Dearing, J. A., Foster, I. D. L., Airey, A., Rippey, B., Wilson, J. B. F. & Appleby, P. G. (1989) Catchment disturbance inferred from paleolimnological

studies of three contrasted sub-humid environments in Morocco. *J. Paleolimnol.* **1**, 293-322.
Foster, I. D. L. (1980) Chemical yields in runoff and denudation in a small arable catchment, East Devon, England. *J. Hydrol.* **47**, 349-368.
Foster, I. D. L. & Dearing, J. A. (1987) Lake-catchments and environmental chemistry: a comparative study of contemporary and historical catchment processes in Midland England. *Geojournal* **14**, 285-297.
Foster, I. D. L., Dearing, J. A., Simpson, A. D. & Carter, A. D. (1985) Lake catchment based studies of erosion and denudation in the Merevale catchment, Warwickshire, UK. *Earth Surf. Proc. Landforms.* **10**, 45-68.
Foster, I. D. L., Dearing, J. A., Airey, A., Flower, R. J. & Rippey, B. (1986a) Sediment sources in a Moroccan lake-catchment: a case study using magnetic measurements. *J. Wat. Resour. Res.* **5**, 320-334.
Foster, I. D. L., Dearing, J. A. & Appleby, P. G. (1986b) Historical trends in catchment sediment yields: a case study in reconstruction from lake sediment records in Warwickshire, UK. *Hydrol. Sci. J.* **31**, 427-443.
Graf, W. L. (1983) The hydraulics of reservoir sedimentation. *Wat. Power Dam Constr.* **35**, 45-52.
Hadley, R. F. (1985) *Long-term monitoring of natural and man-made changes in the hydrological regime and related ecological environments.* Report of IHP-II Project A.3.1. UNESCO, Paris.
Hadley, R. F., Lal, R., Onstad, C. A., Walling, D. E. & Yair, A. (1985) *Recent developments in erosion and sediment yield studies.* Report of IHP-II Project A.1.3.1 UNESCO, Paris.
Hakanson, L. & Jansson, M. (1983) *Principles of Lake Sedimentology.* Springer, Berlin.
Hall, D. G. (1967) The pattern of sediment movement in the River Tyne. In: *Symposium on River Morphology. Reports and Discussions. (General Assembly of Bern, Sept-Oct 1967),* IAHS Publ. no. 75, 117-140.
Heinemann, H. G. (1981) A new sediment trap efficiency curve for small reservoirs. *Wat. Resour. Bull.* 17, 825-30.
Heinemann, H. G. (1984) Reservoir trap efficiency. In: *Erosion and Sediment Yield: Some methods of measurement and modelling.* (ed. R. F. Hadley & D. E. Walling) Geo Books, Norwich, UK.
Hilton, J. (1986) A conceptual framework for predicting the occurrence of sediment focusing and sediment redistribution in small lakes. *Limnol. Oceanogr.* **30**, 1131-1143.
Hilton, J. & Lishman, J. P. (1985) The effect of redox change on the magnetic susceptibility of sediments from a seasonally anoxic lake. *Limnol. Oceanogr.* **30**, 907-909.
Hilton, J., Lishman, J. P. & Allen, P. V. (1986) The dominant process of sediment distribution and focusing in a small, eutrophic, monomictic lake. *Limnol. Oceanogr.* **31**, 125-133.
Kullenberg, B. (1947) The piston core sampler. *Sven. Hydrog. Biolog. Kommis. Skrif. III Hydrografi* 1:2, 1-46.
Lambert, A. M. (1982) Estimation of erosion and sediment yield by volume measurements on a lacustrine river delta. In: *Recent Developments in the Explanation and Prediction of Erosion and Sediment Yield* (Proc. Exeter Symp. July 1984) IAHS Publ no. 137, 171-176.
Lane, E. W. & Koelzer, V. A. (1953) Density of sediments deposited in reservoirs. *Interagency Comm. Measurement Anal. Sed. Loads in Streams* Hydraulic Laboratory, Univ. Iowa. Rept. no. 9.
Ledger, D. C., Lovell, J. P. B. & McDonald, A. T. (1974) Sediment yield studies in upland catchment areas in south-east Scotland. *J. Appl. Ecol.* **11**, 201-206.
Lehman, J. T. (1975) Reconstructing the rate of accumulation of lake sediment: The effect of sediment focusing. *Quatern. Res.* **5**, 541-550.
Likens, G. E. & Davis, M. B. (1975) Post-glacial history of Mirror Lake and its watershed in New Hampshire, USA: an initial report. *Ver. Int. Ver. Limnol.* **19**, 982-993.
Lowe, J. J. & Walker, M. J. C. (1984) *Reconstructing Quaternary Environments.* Longman, London, UK
Mackereth, F. J. H. (1969) A short core sampler for sub aqueous deposits. *Limnol. Oceanogr.* **14**, 145-151.
McManus, J. & Duck, R. W. (1985) Sediment yield estimated from reservoir siltation in the Ochil Hills, Scotland. *Earth Surf. Proc. Landforms* **10**, 193-200.
Oldfield, F. (1977) Lakes and their drainage basins as units of sediment-based ecological study. *Prog. Phys. Geogr.* **3**, 460-504.
Oldfield, F., Battarbee, R. W. & Dearing, J. A. (1983) New approaches to environmental change. *Geogr. J.* **149**, 167-181.
Oldfield, F. & Appleby, P. C. (1984) Empirical testing of ^{210}Pb models for lake sediments. In: *Lake Sediments and Environmental History* (ed. E. Y. Hawarth & J. W. G. Lund), Leicester Univ. Press, Leicester, UK.
Oldfield, F., Worsley, A. T. & Appleby, P. G. (1985) Evidence from lake sediments for recent erosion rates in the highlands of Papua New Guinea. In: *Environmental Change and Tropical Geomorphology* (ed. I. Douglas & T. Spencer), Allen & Unwin, London, UK.

Olsson, I. U. (1986) Radiometric dating. In: *Handbook of Holocene Palaeoecology and Palaeohydrology* (ed. B. Berglund), Wiley, Chichester, UK.

O'Sullivan, P. E. (1983) Annually laminated lake sediments and the study of Quaternary environmental changes - a review. *Quat. Sci. Rev.* **1**, 245-313.

O'Sullivan, P. E., Oldfield, F. & Battarbee, R. W. (1973) Preliminary studies of Lough Neagh sediments. I: Stratigraphy, chronology and pollen analysis. In: *Quaternary Plant Ecology.* (ed. J. Birks & R. G. West) Blackwell, Oxford, UK.

O'Sullivan, P. E., Coard, M. A. & Pickering, D. A. (1982) The use of laminated lake sediments in the estimation and calibration of erosion rates. In: *Recent Developments in the Explanation and Prediction of Erosion and Sediment Yield* (Proc. Exeter Symp. July 1982), 197-207, IAHS Pub. no. 137.

Pemberton, E. L. & Blanton, J. O. (1980) Procedures for monitoring reservoir sedimentation. *Proc. Symp. Surf. Water Impound.*, Minneapolis, Minnesota. 1269-1278.

Pennington, W., Cambray, R. S., Eakins, J. D. & Harkness, D. D. (1976) Radionuclide dating of the recent sediments of Blelham Tarn. *Freshwat. Biol.* **6**, 317-331.

Rausch, D. L. & Heinemann, H. G. (1984) Measurement of reservoir sedimentation. In: *Erosion and Sediment Yield: Some Methods of Measurement and Modelling.* (ed. R. F. Hadley & D. E. Walling). Geobooks, Norwich, UK.

Ritchie, J. C., McHenry, J. R. & Gill, A. C. (1973) Dating recent reservoir sediments. *Limnol Oceanogr.* **18**, 254-263.

Rodda, J. C., Downing, R. A. & Law, F. M. (1976) *Systematic Hydrology.* Newnes-Butterworths, London, UK.

Shalash, S. (1982) Effects of sedimentation on the storage capacity of the High Aswan Dam reservoir. *Hydrobiologia* **92**, 623-639.

Simola, J., Coard, M. A. & O'Sullivan, P. E. (1981) Annual laminations in the sediments of Loe Pool, Cornwall. *Nature*, **290**, 238-241.

Smith, W. O., Vetter, C. P. & Cummings, G. B. (eds.) (1960) Comprehensive study of sedimentation in Lake Mead, 1948-49. *USGS Prof. Pap.* 265.

Stromquist, L. (1981) Recent studies on soil erosion, sediment transport and reservoir sedimentation in semi-arid central Tanzania. In: *Tropical Agricultural Hydrology* (ed. R. Lal & E. W. Russell) Wiley, Chichester.

Szechowycz, R. W. (1973) Sedimentation in Mangla Reservoir. *J. Hydraul. Div. ASCE* **HY9**, 1551-1572.

Thompson, R. (1986) Palaeomagnetic dating. In: *Handbook of Holocene Palaeoecology and Palaeohydrology* (ed. B. Berglund) Wiley, Chichester, UK.

Thompson, R., Bradshaw, R. H. W. & Whitely, J. E. (1986) The distribution of ash in Icelandic lake sediments and the relative importance of mixing and erosional processes. *J. Quat. Sci.* **1**, 3-11l.

Thompson, R. & Oldfield, F. (1986) *Environmental Magnetism.* Allen &Unwin, London. UK.

Tolonen, K., Siiriainen, A. & Thompson, R. (1975) Prehistoric field erosion sediment in Lake Lojarv, S. Finland and its palaeomagnetic dating. *Ann. Bot. Fennic.* **12**, 161-164.

Trujillo, L. F. (1982) Trap efficiency study, Highland Creek flood retarding reservoir near Kelseyville, California, water years 1966-77. *USGS Wat. Supply Pap.* 2182.

Vanoni, V. A. (1975) (ed). *Sedimentation Engineering.* ASCE. Man. Rep. Engng Pract. no. 54.

Walling, D. E. (1978) Reliability considerations in the evaluation and analysis of river loads. *Z. Geomorphol. Suppl. Bd.* **29**, 29-42.

Walling, D. E. (1983) The sediment delivery problem. *J. Hydrol.* **65**, 209-37.

Walling, D. E. & Webb, B. W. (1981) The reliability of suspended sediment load data. In: *Erosion and Sediment Transport Measurement* (Proc. Florence Symp. June 1981), 177-194. IAHS Publ. no. 133.

Wolman, M. G. (1967) A cycle of sedimentation and erosion in urban river channels. *Geografiska Annal.* **40A**, 385-395.

Wright, H. E. (1980) Cores of soft lake sediments. *Boreas* **9**, 107-114.

Yamomoto, A. (1984) Grain size variation. In: *Lake Biwa.* (ed. S. Horie). Dr W. Junk, The Hague, The Netherlands.

Young, A. (1958) A record of the rate of erosion on millstone grit. *Proc. Yorks. Geol. Soc.* **31**, 149-156.

Erosion, Transport and Deposition Processes (Proceedings of the Jerusalem Workshop, March-April 1987). IAHS Publ. no. 189, 1990.

Empirical relationships for the transport capacity of overland flow

GERARD GOVERS

Research Associate, National Fund for Scientific Research, Belgium, Laboratory of Experimental Geomorphology Catholic University of Leuven, Redingenstratt 16B, 3000 Leuven, Belgium

Abstract This paper reports the results of an experimental study on the transport capacity of overland flow. Experiments were carried out with five materials varying from silt to coarse sand, using a wide range of discharges and slopes, in a flume with a plane bed. The experiments provided the necessary data to establish highly correlated relationships between the sediment transport capacity of the flow and different hydraulic parameters, which are a necessary element of all physically-based erosion models. Some important characteristics of these relationships are identified. These include the limited range of validity of individual relationships and the apparent lack of connection between initiation of sediment motion and sediment transport. Some relationships can also be applied to rough surfaces and to surfaces with a vegetation cover. It is believed that the proposed set of empirical relationships can be of use in the study of many aspects of soil erosion and deposition.

INTRODUCTION

The sediment transporting capacity of overland flow is a parameter of fundamental importance in the physically-based description of soil erosion and deposition processes. The transport capacity of the flow at a given point equals the maximum net erosion potential upslope. Flow incision, for example in a tractor wheeling, will only occur when the transport capacity of the flow is sufficiently high to evacuate all the material that is transported into the flowpath from the inter-rill areas. Knowledge of the amount of sediment that can be transported over a concave, basal slope segment (where sedimentation takes place) will often be of crucial importance when the response of a first-order watercourse has to be understood and predicted. The variable transportability of grains of different size and density can cause size sorting during erosion and transport. Furthermore, some researchers state that the detachment capacity of the flow can be directly related to its transport capacity. Foster & Meyer (1972a) express this relationship as:

$$D_r/D_c + q_s/T_c = 1 \qquad (1)$$

where:

D_r = the detachment rate per unit of time and per unit surface of the bed;
D_c = the detachment rate for clear water flow;
q_s = the unit solid discharge; and
T_c = the transport capacity of the flow.

Foster (1982) transforms this relationship into:

$$D_r = \alpha \, (T_c - q_s) \qquad (2)$$

where:

$$\alpha = D_c / T_c$$

Equation (2) describes the detachment rate in terms of the difference between transport capacity and actual transport rate. It implies that the detachment rate will decline as the transport capacity is approached.

Almost all physically-based erosion models that have been developed during the last two decades contain a sediment transport capacity equation. In most cases, a formula that has been developed for rivers is used, although the empirical coefficients are sometimes modified. Foster & Meyer (1972b) proposed the formula of Yalin (1963) as being the most applicable to shallow flow conditions. This formula has subsequently been used by several other modellers (Dillahah & Beasley, 1983; Kahnbilvardi *et al.*, 1983; Park *et al.*, 1982). Savat (1979) found that sediment concentrations measured during recirculating flume experiments on a loamy soil were generally only about 25% of those predicted by the Yalin formula. Alonso *et al.*, (1981) concluded that the formula gave good results for sheet flow on concave surfaces with relatively low sediment loads. Preliminary results obtained by the author showed that the transport capacity of overland flow on steep slopes could be much higher than predicted by the expression of Yalin (Govers, 1985).

Other formulae that have been employed include the Ackers & White (1973) formula used by Morgan (1980); the Bagnold (1966) formula used by Rose *et al.* (1983) and Gilley *et al.* (1985); the Kalinske (1942) formula used by Komura (1976) and Mossaad & Wu (1984); and the Yang (1973) formula used by Wilson *et al.* (1984).

A different approach has been proposed by Tödten (1976) and Prasad & Singh (1982). They derive complex transport expressions from basic physical considerations of sediment detachment and movement. Other modellers relate transport capacity directly with a simple hydraulic parameter such as shear stress (David & Beer, 1975; Croley, 1982) or stream power (Kirkby, 1980).

The great variety of formulae that have been used in theoretical approaches is largely due to the fact that insufficient experimental data are available to test the validity of the proposed equations. Nevertheless, it cannot be expected that any transport formula can be successfully applied without suitable calibration over the whole range of field conditions. Even the more sophisticated river formulae (e.g. Yalin, 1963) contain one or more empirically determined constants. Calibration requires the availability of an experimental data set acquired under controlled conditions.

In this paper experimental results on the transport capacity of overland flow are presented. Experiments were carried out on slopes ranging from 1 to 12° and with unit discharges between 2 and 100 $cm^3\ cm^{-1}\ s^{-1}$. The results were analysed, not by comparing them with predictions based on existing formulae, but by relating them to hydraulic parameters which are commonly considered to be relevant to the transport capacity of river flow. This approach has the advantage of leading to simple expressions, which are easy to manipulate, whilst the best correlation that can be obtained might be better than that associated with an existing formula.

EXPERIMENTAL PROCEDURE

All experiments were carried out in a flume 12 m long and 0.117 m wide of which only 6 m was effectively used, in order to avoid water surface instability (Fig. 1). The bottom of the flume was covered with a 1 cm thick layer of sediment that was carefully smoothed. Water was then applied and, when equilibrium was established, water and sediment were collected at the lower end of the flume during a short time interval.

This experimental procedure is clearly different from those used to determine the transport capacity of river flows. These involve either recirculating sediment and water over a movable bed until equilibrium is established (e.g. Rathbun & Guy, 1967), or adding sediment to clear water flowing over a moveable bed and varying the sediment supply rate until stable conditions are obtained (e.g. Luque & Van Beek, 1976). These procedures were not feasible because of the high sediment concentrations that can be transported by overland flow on steeper slopes. The only disadvantage of the method described above is that in some cases it might be possible that sediment transport capacity is not reached within a distance of 6 m. However, experiments carried out with an effective length of only 3 m yielded comparable results (Fig. 2). It can therefore be concluded that a length of 6 m is sufficient to reach full transport capacity.

Experiments were carried out using five well-sorted quartz materials with a median grain size varying from silt to coarse sand (Table 1). Slopes were 1, 2, 5, 8 and 12° and unit discharges from 2 to 100 $cm^3\ cm^{-1}\ s^{-1}$. In total, 436 measuring runs were undertaken.

The hydraulic characteristics of such flows can be accurately calculated using an algorithm developed by Savat (1980), providing no sediment is present. In his paper, Savat (1980) takes account of sediment load by increasing water viscosity, which leads to a lower Reynolds flow number and therefore to a reduction of mean velocity and to an increase of water depth.

A number of experiments were carried out to investigate the validity of this approach. Comparison of measured with predicted velocities revealed that, instead of a velocity decrease, a significant velocity increase (up to 40%) occurs (Fig. 3). The fact that the presence of suspended sediment increases the flow velocity in rivers has been known for a long time (e.g. Vanoni & Nomicos, 1959). Two mechanisms are considered to be responsible for this. The presence of sediment adds momentum to the flow and alters the turbulence structure. In

Fig. 1 *View of the experimental flume.*

earlier studies it was believed that the von Karman constant was reduced, but recent research stresses the impact of the sediment load on the wake term of the turbulent flow equation, such that the von Karman constant remains basically unaffected (Coleman, 1981, Parker & Coleman, 1986).

The models developed to describe turbulent flow carrying suspended sediment are not directly applicable to the situation considered here. This is especially the case for the coarsest material where the water film is often only a few grain diameters thick, so that one cannot speak of true suspension. Furthermore, sediment concentrations are so high that grain interactions become important, leading to the development of significant dispersive and tangential stresses (Bagnold, 1954). It can therefore be expected that the influence of grain size is less than when true suspension is being considered (Parker & Coleman, 1986).

In order to estimate mean velocity for our experimental results, the mean velocity was calculated according to the procedure of Savat (without taking into account the influence of sediment on viscosity). This estimate was

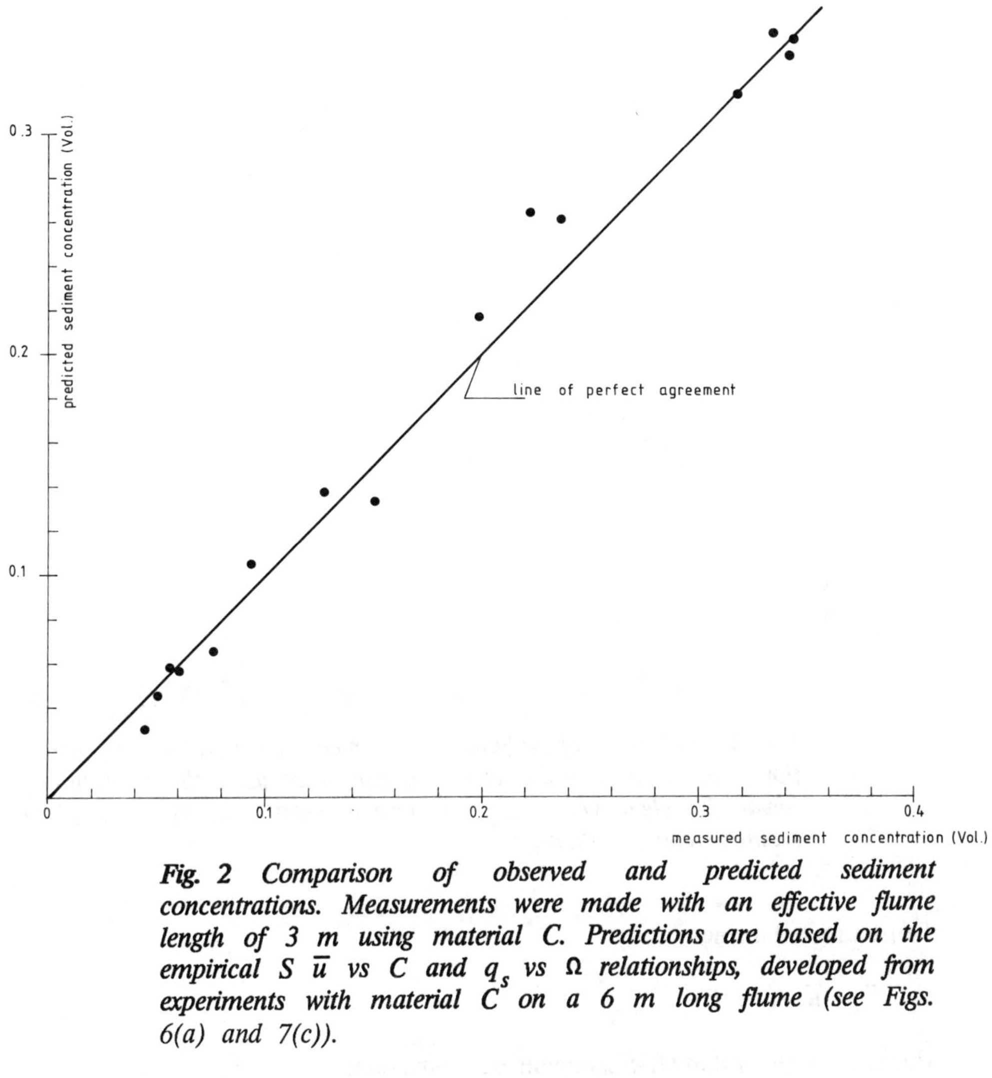

Fig. 2 *Comparison of observed and predicted sediment concentrations. Measurements were made with an effective flume length of 3 m using material C. Predictions are based on the empirical* $S\,\bar{u}$ *vs C and* q_s *vs* Ω *relationships, developed from experiments with material C on a 6 m long flume (see Figs. 6(a) and 7(c)).*

Table 1 *Characteristics of the materials used in the experiments (CSF = Corey shape factor =* $c(ab)^{-1/2}$*, where c = the shortest axis, b = the intermediate axis and a = the longest axis of the grain)*

Material	D_{50} *(µm)*	*CSF*
A	*58*	*0.59*
B	*127*	*0.79*
C	*218*	*0.71*
D	*414*	*0.64*
E	*1098*	*0.66*

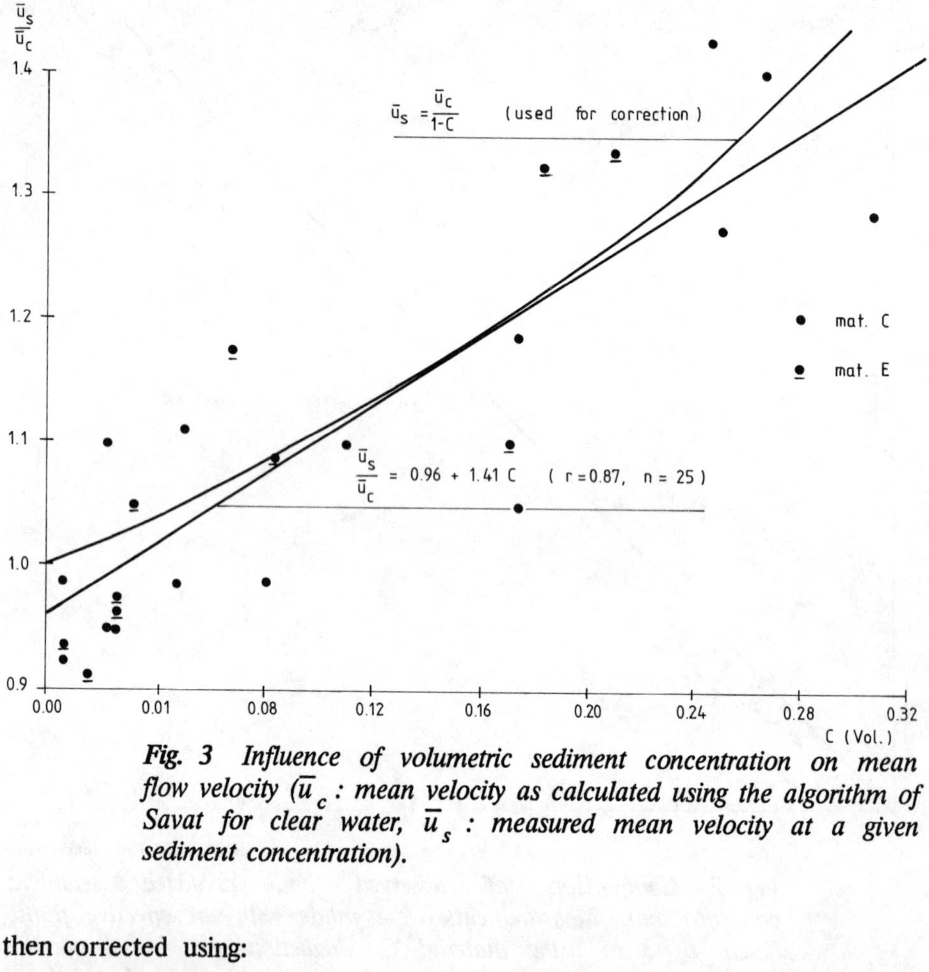

Fig. 3 Influence of volumetric sediment concentration on mean flow velocity ($\bar{u}_c$: mean velocity as calculated using the algorithm of Savat for clear water, $\bar{u}_s$: measured mean velocity at a given sediment concentration).

then corrected using:

$$\bar{u}_s = \bar{u}_c / (1 - C) \quad (3)$$

where: C = the volumetric sediment concentration;
$\bar{u}_c$ = the velocity calculated for clear water flow; and
$\bar{u}_s$ = the actual velocity of the sediment-laden water flow.

Hydraulic parameters were then calculated without taking into account the influence of the sediment on the specific weight of the fluid and fluid depth, because no information is at present available on grain velocities in overland flow. As a consequence the amount of sediment present per unit surface of the bed and therefore total shear stress and energy dissipation cannot be calculated.

ANALYSIS OF RESULTS

Relevant hydraulic parameters

Most of the older river transport formulae rely on the concept of excess shear. Sediment transport capacity (expressed as a solid discharge) is then related to the excess shear stress, being the difference between the actual shear stress and the critical shear stress necessary to initiate movement. The shear stress is calculated as:

$$\Gamma = \rho \, g \, R \, S \tag{4}$$

where :

ρ = the density of the fluid;
g = the gravitational acceleration;
R = the fluid depth; and
S = the slope.

Bagnold (1966) was the first to introduce an equation which was no longer based on a balance of forces, but on a balance of energy. He introduced the concept of stream power, which represents the amount of energy dissipated per unit of time and per unit of bed surface. The stream power can therefore be expressed as the product of the shear stress and the mean flow velocity. Later, he stated that there is only a unique relationship between sediment transport capacity and stream power if the depth of the flow is constant. Finally he proposed the following relationship (Bagnold, 1977, 1980):

$$q_s \sim (w - w_{cr})^{1.5} / (R^{2/3} D^{1/2}) \tag{5}$$

where :

w = the stream power; and
w_{cr} = the critical stream power value at which sediment movement starts.

The expression $(w - w_{cr})^{1.5}/R^{2/3}$ might be considered to be an effective stream power (Ω), corrected for the influence of depth.

Yang (1972) introduced the concept of unit stream power, which is the amount of energy dissipated per unit time and per unit weight of the flow and which is equal to the product of slope and mean velocity. This quantity should not be related to solid discharge, but to sediment concentration, so:

$$\log C = A + B \log(S \, \bar{u}) \tag{6}$$

where :

$\bar{u}$ = the mean flow velocity; and
C = the concentration by weight, expressed in ppm.

Later, he developed more complex dimensionless equations, which are based on four dimensionless groups describing the flow and the sediment: one equation does take into account the existence of a critical stream power value required to initiate motion, while the other does not (Yang, 1973). The most important parameter is, according to Yang, the dimensionless unit stream power, which equals the stream power divided by the fall velocity of the particles. Recently, Moore & Burch (1986) concluded that the formula of Yang (1973) gave promising results with respect to the transport capacity of sheet and rill flow.

In the following sections, our results are related to the parameters described above. Logarithms were used in the relationships involving solid discharge because of the wide range of absolute values.

Shear stress

A reasonable relationship between solid discharge and shear stress was found for all tested grain sizes (Fig. 4). The curves for materials A and B show a well defined break at a shear stress of about 20 g cm^{-1} s^{-2}. Apart from this observation, all relationships are more or less linear. No clear tendency towards a vertical asymptote is present, which indicates that it is not necessary to introduce a critical shear stress into the relationship.

The slope of the regression equation decreases with increasing grain size. For the finest materials, the shear stress coefficient is greater than 4.0, while it approaches 2.5 for material E. More surprising is the fact that the intercept increases with grain size, so that at low shear stresses, the transport capacity of overland flow is higher for coarse than for fine sediment (Fig. 5).

The effective stream power

For materials C, D and E, solid discharge is well related to the effective stream power as derived by Bagnold (1980) (Fig. 6). For the finest materials, the relationship was much less satisfactory. Again, the relationships are more or less linear over the whole range. The influence of grain size is thus clearly different from that supposed by Bagnold (1980), who stated that sediment transport capacity should be inversely related to the square root of the grain diameter. The effective stream power coefficient is not constant but decreases with grain size. Furthermore, at low effective stream power values, sediment transport capacity increases with grain size.

The unit stream power

For the finest sediments a very good relationship exists between volumetric sediment concentration and unit stream power, at least if the data obtained on a 12° slope are excluded (Fig. 7). On this slope, sediment concentrations were considerably higher than on lower slopes at comparable unit stream power values, especially for low discharges. It should also be noted that a limiting volumetric concentration of about 0.32 was reached at high unit stream power values. A further increase of unit stream power did not cause an increase of sediment concentration. The unit stream power exponent increases with increasing grain size. Also, a critical value had to be introduced into the relationships, which appeared to be more or less independent of the grain size ($S\bar{u}_{cr}$ = 0.4 cm^{-1}).

Over the whole range covered by the equations, there is an increase of

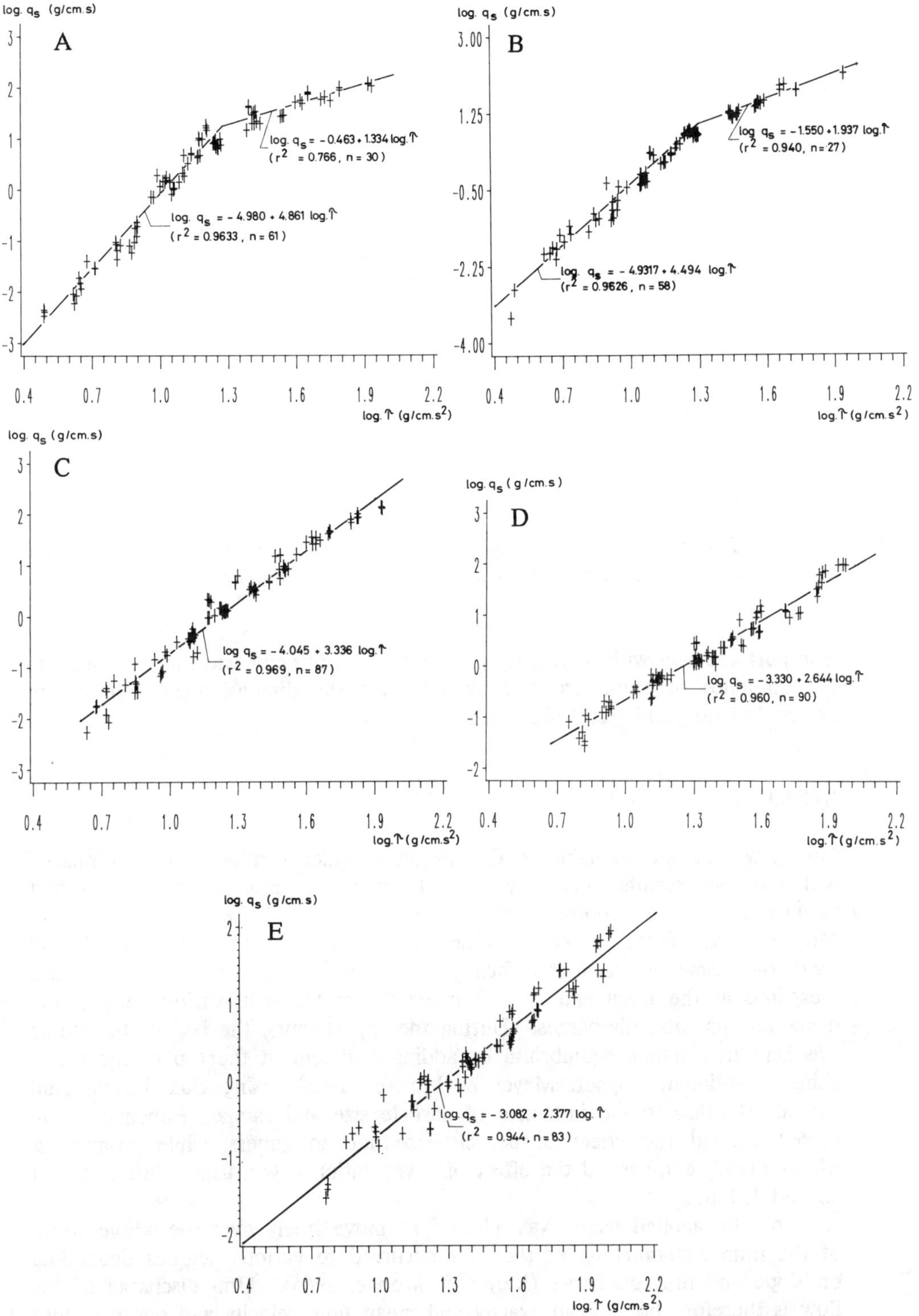

***Fig. 4** Fluid shear stress vs unit solid discharge relationships for all materials A-E.*

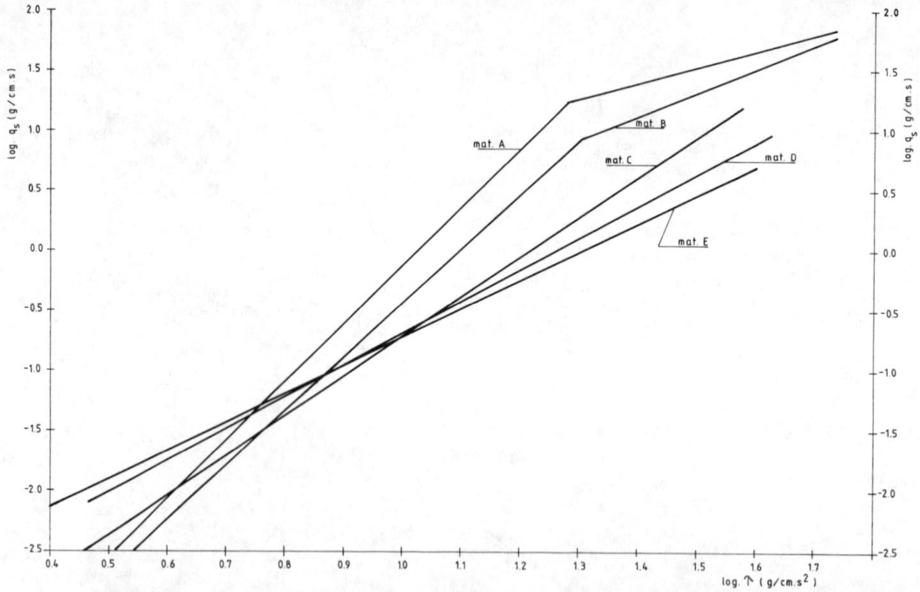

Fig. 5 *Comparison of regression equations relating unit solid discharge with fluid shear stress obtained for various materials (A-E).*

transport capacity with increasing grain size. An attempt was made to include the influence of grain size directly, by using the dimensionless unit stream power, but no good result was obtained.

Validation

Only few data are available in the literature which can be readily compared with our own results. Probably the data which are most suitable for a first verification of the proposed relationships are those collected by Meyer & Monke (1965), Bubenzer *et al.* (1966) and Kramer & Meyer (1969), who all used the same experimental facility. Their papers discuss sediment loads measured at the basal end of a 2 m by 0.6 m flume at various slopes and using various total discharges. During the experiments, the bed of the flume was kept in dynamic equilibrium by adding sediment at the top of the flume using a sediment hopper. Meyer & Monke (1965), using glass beads, paid special attention to the influence of particle size and rainfall. Bubenzer *et al.* (1966) studied the effect of particle size and roughness, while Kramer & Meyer (1969) emphasised the effect of a vegetation cover, using glass beads of 33 and 121 μm.

As the applied water was allowed to move freely over the whole width of the flume, channelling of the flow occurred to various degrees depending on slope and unit discharge (Meyer & Monke, 1965). Unit discharge of the flow is therefore not known exactly and mean flow velocity and depth cannot be accurately calculated. However, Kramer & Meyer (1969) report mean flow velocities, so that for their data mean unit discharge and flow depth could be

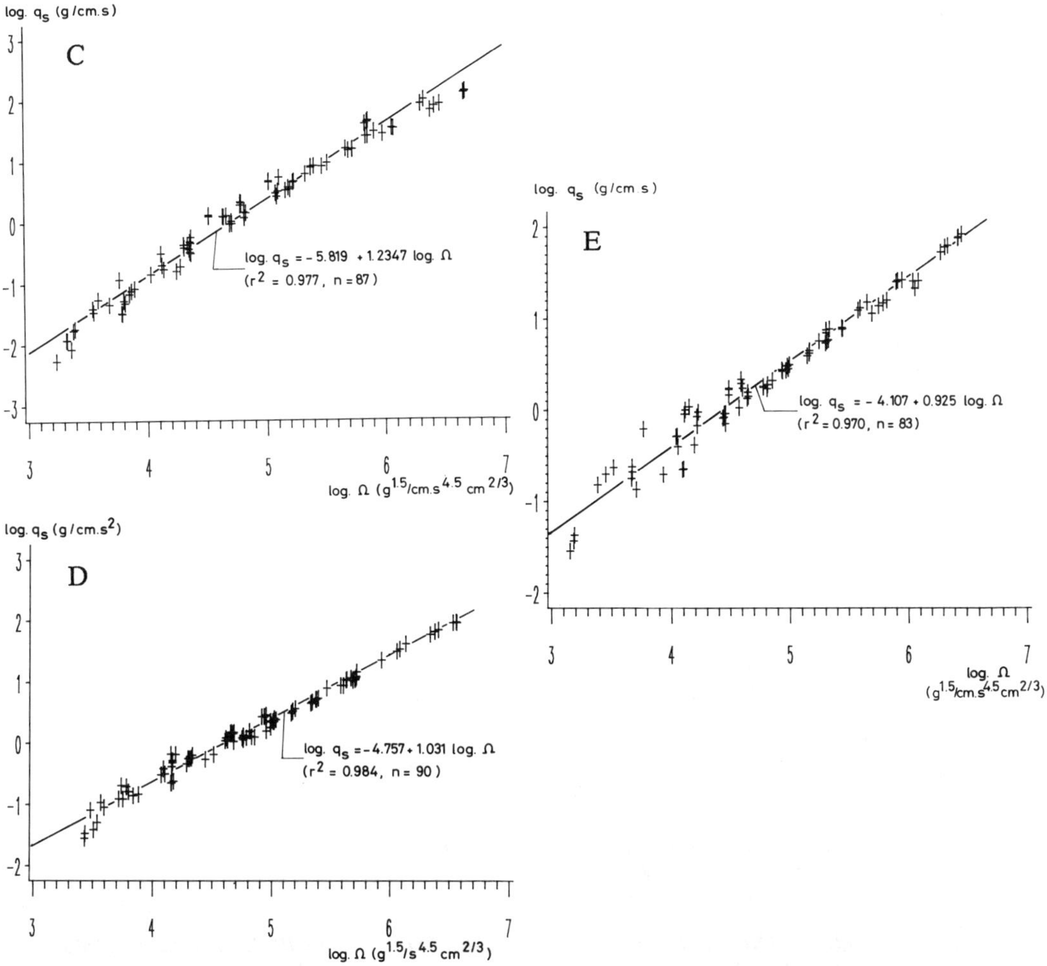

Fig. 6 Effective stream power vs unit solid discharge relationships for materials C, D and E.

calculated using the procedure of Savat (1980) (see Govers & Rauws, 1986). Correction factors were then calculated from their data, allowing the estimation of unit discharge and the calculation of estimated mean velocity and depth for the data of Meyer & Monke (1965) and Bubenzer *et al.* (1966). Total discharges were always low, so that the flow was in all cases laminar or nearly laminar. Predicted sediment loads were then calculated using these estimates and deriving appropriate constants for the various equations from Fig. 8.

It appeared that the sediment loads of very low energy flows ($S\,\overline{u} < 0.7$ cm s^{-1}) measured by these authors were considerably lower than the sediment transport capacity predicted by the various empirical equations presented in this paper. This was especially true for coarse materials ($D_{50} > 450$ μm). However, if these data are excluded, there appears to be a good agreement (taking into account the rather approximate estimates of hydraulic charac-

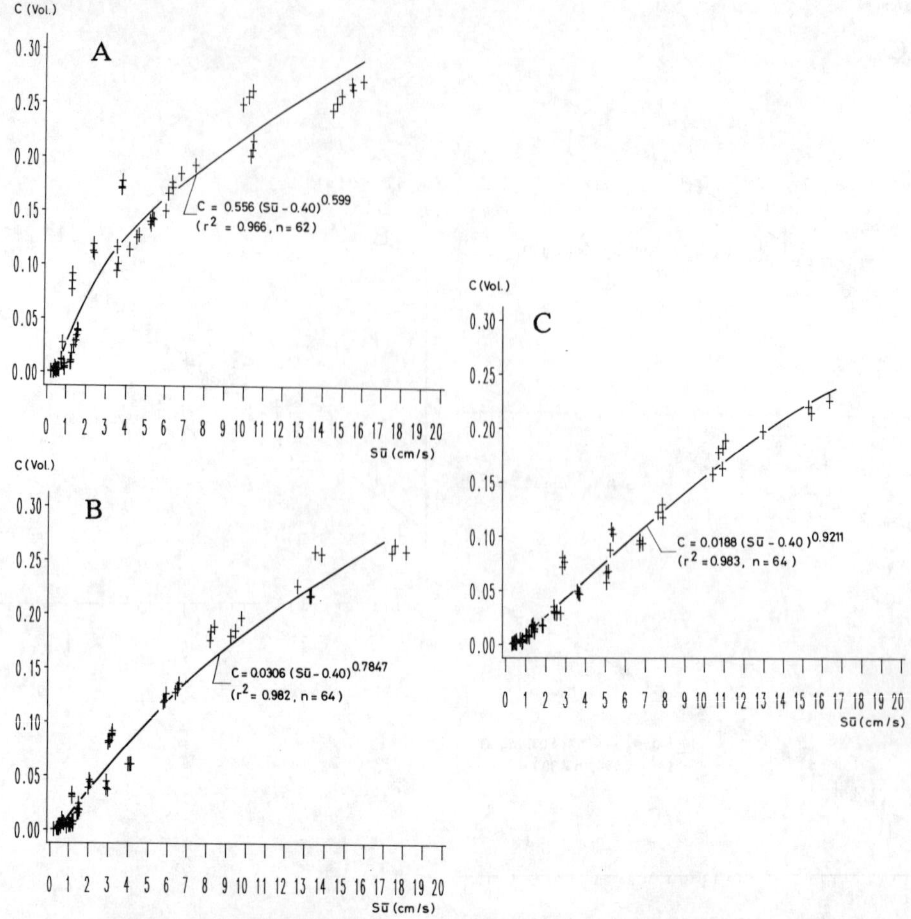

***Fig. 7** Unit stream power vs volumetric concentration relationships for materials A, B and C.*

teristics) between the predicted and the actually measured sediment load (Fig. 9). The effective stream power gives the best results, whilst scatter is greater for the shear stress and unit stream power relationships. A possible explanation for the discrepancy at low energy values will be discussed later. It should also be stressed that the transport capacity of overland flow in this range is always rather low.

DISCUSSION

The results obtained clearly indicate that it is indeed possible to predict the transport capacity of overland flow using simple hydraulic parameters. However, the effective stream power and the unit stream power can only successfully be applied within a given grain size range.

Relationships other than those presented in this paper may also have a good predictive capacity. Sediment transport capacity, expressed as a concen-

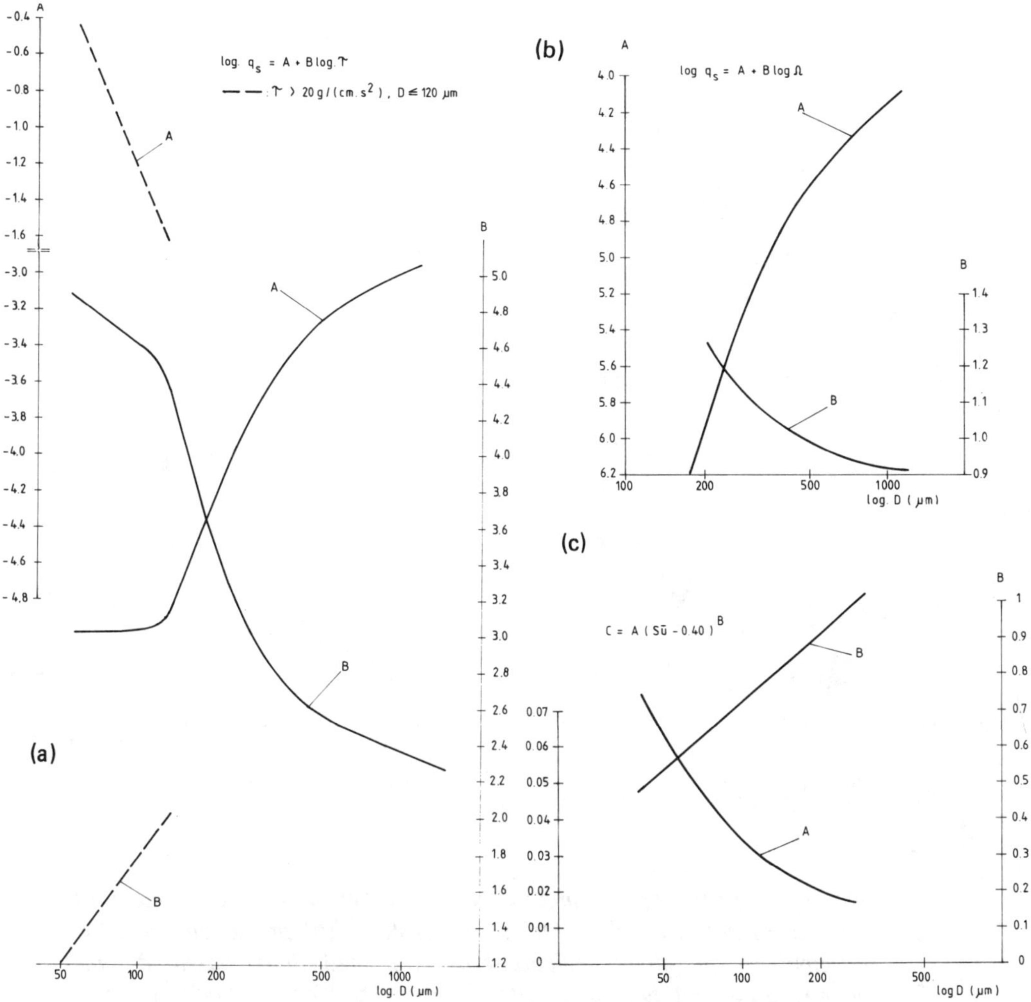

Fig. 8 Nomographs to determine the coefficients of transport capacity relationships as a function of grain size: (a) shear stress, (b) effective stream power, (c) unit stream power.

tration, is, in the laminar to transitional flow range, also well related to the shear velocity of the flow, due to the fact that in the laminar flow range unit stream power and shear velocity are uniquely related (Govers & Rauws, 1986). These relationships always indicate a sharp rise in sediment transport capacity from a shear velocity of about 3 cm s^{-1} (Fig. 10). This shear velocity value is therefore a valuable threshold for rill initiation, providing soil mechanical resistance is not too important (Govers, 1985; Rauws, 1987).

The shear stress, effective stream power and unit stream power relationships can, for a given grain size, be used in the turbulent as well as in the laminar flow range, despite important differences in flow structure and velocity profile. This contrasts with the findings on initial bed instability in overland flow. Critical shear stresses to initiate motion are clearly higher in

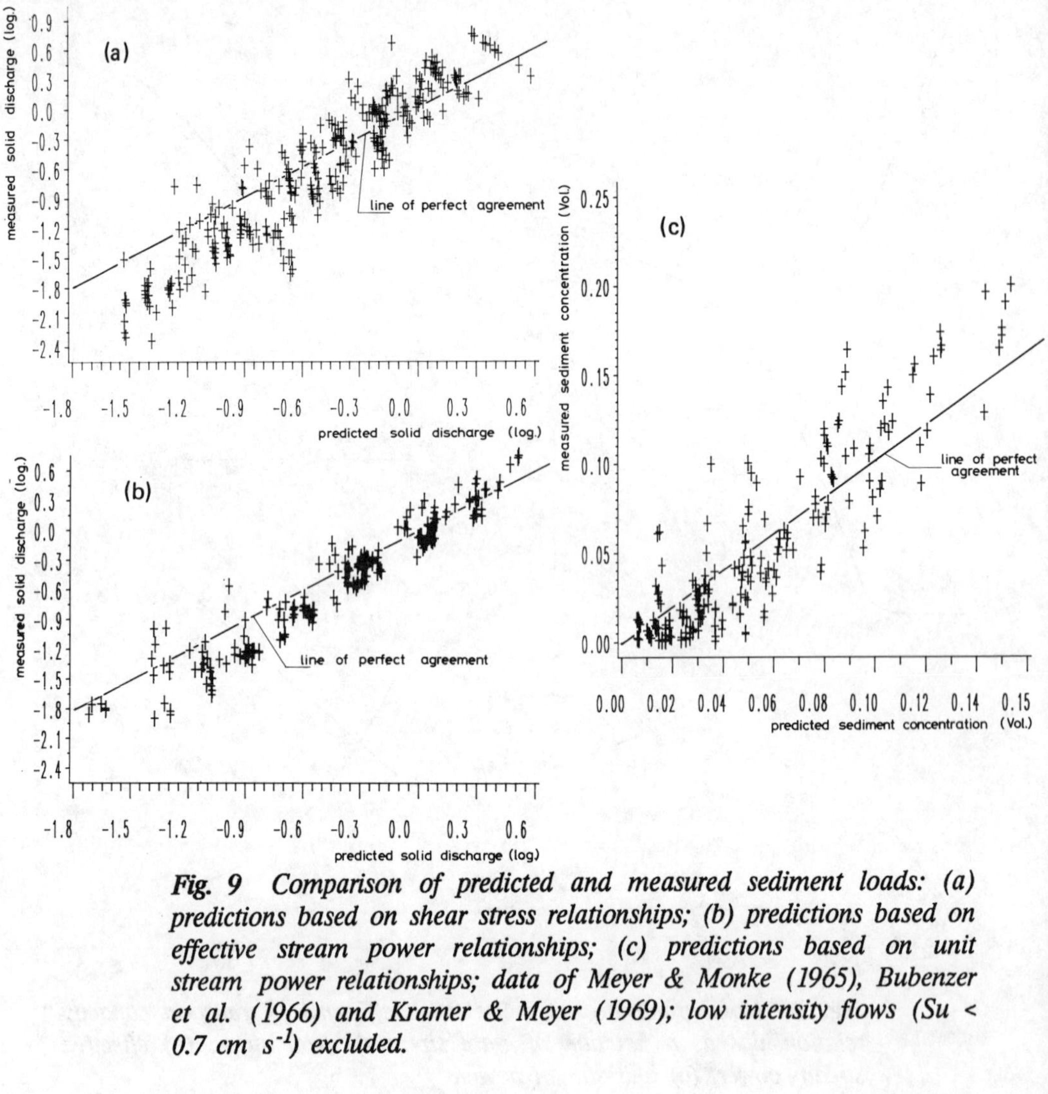

Fig. 9 Comparison of predicted and measured sediment loads: (a) predictions based on shear stress relationships; (b) predictions based on effective stream power relationships; (c) predictions based on unit stream power relationships; data of Meyer & Monke (1965), Bubenzer et al. (1966) and Kramer & Meyer (1969); low intensity flows (Su < 0.7 cm s^{-1}) excluded.

laminar flows and increase with increasing grain size, at least when loose, homogeneous beds are compared (Govers, 1987). However, transport capacity increases with grain size at low flow intensity values. Furthermore, it appears that measurable amounts of coarse sediment are transported at shear stress values below the experimentally determined threshold value for laminar flows.

The latter can be explained by the influence of flume length. During the experiments on the initiation of motion it could be observed that, when coarse particles are set in motion in a laminar flow, they keep moving steadily and mobilize other grains by collision. This will result in a net increase of transport rate with distance from the upper flume end up to an unknown length. This phenomenon becomes more and more important with increasing grain size. A considerable amount of sediment can then be expected to be transported, although the number of primary detachments caused solely by the

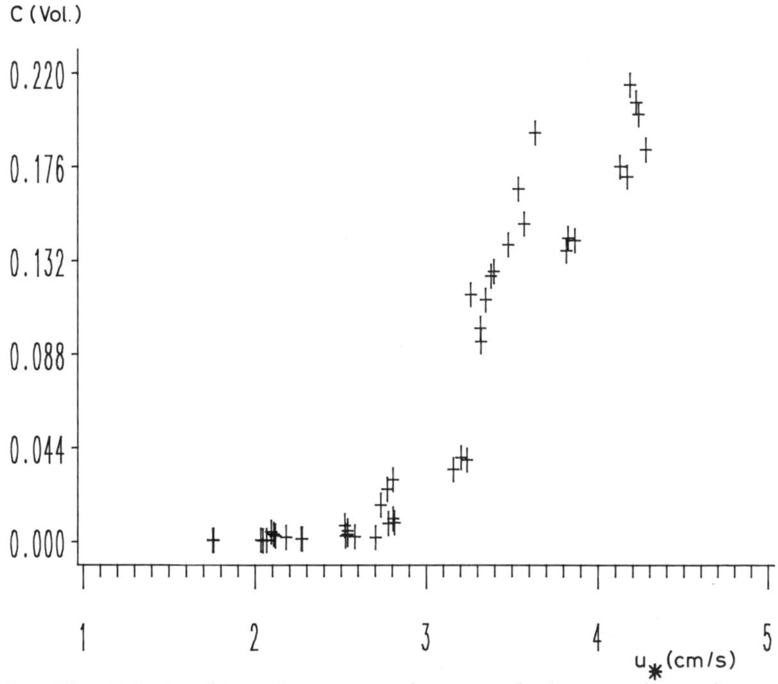

***Fig. 10** Relationship between shear velocity and volumetric sediment concentration (material A, Reynolds flow number < 1800).*

fluid is very low. This process might also explain the discrepancies between our results and those of the American researchers in the low energy range. They used a flume only 2.0 m in length, so that, when grain collision is the primary agent of detachment, full transport capacity might not have been reached.

Another factor that needs to be considered is grain velocity. From theoretical considerations it can readily be shown that the velocity at the center of a grain top in contact with the bed in a laminar flow is proportional to the grain size. It can therefore be expected that, as long as grains are transported near the bed, the coarsest grains will move the fastest. This was experimentally verified by Parsons (1972) for glass beads and sand grains moving over a smooth bed. As the unit solid discharge equals the product of the mass of sediment in motion per unit surface of the bed and the mean grain velocity, it is then acceptable that transporting capacity increases with grain size, as long as sediment transport takes place in laminar flow or in a laminar sublayer.

The fact that sediment transport in overland flow conditions can be predicted by the same hydraulic parameters which are also considered to be of fundamental importance with respect to sediment transport in rivers, does not mean that river formulae can be directly applied to overland flow conditions. This is already evident from the ambivalent effect of grain size. As another example, it may be mentioned that the unmodified equation of Yang, which does not take into consideration a critical unit stream power, only yields a correlation coefficient of 0.38, when it is applied to all data. If a multiple regression is carried out, so that the basic form of the Yang equation is maintained, while the empirical constants are allowed to vary, a correlation coefficient of 0.90 is achieved.

The above described observations imply that in a relationship of the form:

$$Tc = A\, q^b\, S^c \tag{7}$$

where: q = the unit discharge ($cm^3\ cm^{-1}s^{-1}$), the proportionality factor as well as the slope and discharge exponent will vary with grain size and flow type. Coefficients resulting from regression analysis (in logarithmic form) on our data are presented in Table 2. In the laminar range the discharge

Table 2 Regression coefficients of the equation: $\log q_s = A + B\log q + C\log S$; obtained for various grain sizes and distinguishing between laminar and turbulent flow

	Laminar					Turbulent				
Material	*A*	*B*	*C*	r^2	*n*	*A*	*B*	*C*	r^2	*n*
A	1.56	1.65	2.62	0.98	29	0.24	1.66	1.44	0.87	61
B	1.58	1.55	2.76	0.98	28	0.17	1.80	1.69	0.95	57
C	1.21	1.70	2.50	0.98	31	0.74	1.50	1.96	0.98	56
D	0.79	1.53	1.97	0.96	20	0.76	1.24	1.71	0.99	69
E	0.51	1.73	1.76	0.98	21	0.85	1.04	1.47	0.97	62

coefficient appears to be more or less constant, while the slope coefficient decreases with increasing grain size. In contrast, the discharge coefficient decreases with increasing grain size for turbulent flow, while the slope coefficient reaches a maximum value for material C. It is interesting to note that the discharge coefficient always exceeds one. If it is assumed that unit discharge increases linearly with the distance from the divide and that the detachment rate at a given point is proportional to the increase in transport capacity, then detachment rate will always increase with distance from the divide. Dynamic equilibrium will therefore only be possible on a concave slope (Carson & Kirkby, 1972). The q_s versus S, q relationships discussed above are only valid on plane beds. The use of more relevant hydraulic parameters has the advantage that the validity of the relationships can be extended to irregular beds, which are much more common in nature.

It was shown in a former paper that the relationship between the sediment concentration that can be transported and the unit stream power does not appear to be fundamentally modified when the velocity of the water is reduced due to additional friction caused by bed surface irregularities or vegetation elements (Govers & Rauws, 1986). The reduction of sediment transport capacity seems to be directly related to the reduction in energy dissipation.

The effective stream power is an empirical parameter, for which it is difficult to assess the appropriate value on irregular surfaces. However, for material B a reduction of the flow velocity to 50% of the original value will, according to the unit stream power equation, result in a reduction of transport capacity to about 53% of the original value (if the unit stream power is much greater than the critical value). When the effective stream power relation is applied, the same velocity reduction yields a transport capacity of about 56% of the original value (assuming an infinitely wide flow). This may be an indication that the effective stream power relationships are not fundamentally changed by bed irregularities.

With respect to the shear stress, computational procedures exist to split up total shear stress into a grain component, representing that part of the shear stress which can effectively be used for sediment transport, and a form component, which is that part of the shear stress that is dissipated on major surface irregularities. A modification of the procedure of Einstein & Barbarossa (1952) has already been used to predict sediment transport capacity and rill generation (Govers & Rauws 1986, Rauws & Govers, 1988). This computational method might yield considerably higher reductions in sediment load, as the grain shear stress is proportional to the square of the mean velocity, although this is partly compensated by the introduction of an appropriate grain roughness.

CONCLUSIONS

The transport capacity of overland flow was investigated experimentally and it has been shown that it is possible to predict sediment transport capacity using simple empirical relationships, which show considerable variation with grain size, but not with flow regime. Calculation of mean flow velocity and depth should take into account the presence of sediment load which can increase the velocity by up to 40%. The proposed relationships were validated using the results of US researchers, who used a different experimental set-up. It appeared that there is a good agreement between predicted and measured load, at least when very low energy flows were excluded.

It is believed that the proposed relationships can contribute significantly to the operationalization of physically-based erosion models. However, more information should be collected concerning the influence of sediment specific density and of bed surface irregularities. Furthermore, the applicability of the proposed relationships could be considerably extended, when data concerning the interaction between transport capacity, actual load and detachment rate become available.

REFERENCES

Ackers, P. & White, W. R. (1973) Sediment transport: new approach and analysis. *J. Hydraul. Div. ASCE* **99**, 2041-2060.

Alonso, C. V., Neibling, W. H. & Foster, G. R. (1981) Estimating sediment transport capacity in watershed modelling. *Trans. Am. Soc. Agric. Engrs* **24**, 1211-1221.

Bagnold, R. A. (1954) Experiments on gravity-free dispersion of large solid spheres in a Newtonian fluid under shear. *Proc. Roy. Soc.* Ser. A, **225**, 49-63.
Bagnold, R. A. (1966) An approach to the sediment transport problem from general physics. *USGS Prof. Pap.* **422**.
Bagnold, R. A. (1977) Bedload transport by natural rivers. *Wat. Resour. Res.* **31**, 303-311.
Bagnold, R. A. (1980) An empirical correlation of the bedload transport rates in flumes and natural rivers. *Proc. Roy. Soc.* Ser. A, **372**, 453-473.
Bubenzer, G. D., Meyer, L. D. & Monke, E. G. (1966) Effects of particle roughness on soil erosion by overland flow. *Trans. Am. Soc. Agric. Engrs.* **4**, 562-564.
Carson, M. A. & Kirkby, M. J. (1972) *Hillslope Form and Process.* Cambridge University Press, Cambridge, UK.
Coleman, N. L. (1981) Velocity profiles with suspended sediment. *J. Hydraul. Res.* **19**, 211-229.
Croley, T. E. (1982) Unsteady overland sedimentation. *J. Hydrol.* **56**, 325-346.
David, W. P. & Beer, C. E. (1975) Simulation of soil erosion I. Development of a mathematical model. *Trans. Am. Soc. Agric. Engrs.* **18**, 126-129.
Dillaha, T. A. & Beasley, D. B. (1983) Distributed parameter modelling of sediment movement and particle size distribution. *Trans. Am. Soc. Agric. Engrs.* **26**, 1766-1773.
Einstein, H. A. & Barbarossa, N. L. (1952) River channel roughnes. *Trans. ASCE* **117**, 1121-1132.
Foster, G. R. (1982) Modelling the erosion process. In: *Hydrologic Modelling of Small Watersheds* (ed. C. T. Hann), American Society of Agricultural Engineers, St. Joseph, 297-370.
Foster, G. R. & Meyer, L. D. (1972a) A closed-form equation for upland areas. In *Sedimentation, Symp. to Honour Prof. H. A. Einstein,* Fort Collins, (ed. H. Shen), 12.1-12.17.
Foster, G. R. & Meyer, L. D. (1972b) Transport of soil particles by shallow flow. *Trans Am. Soc. Agric. Engrs.* **15**, 99-102.
Gilley, J. E., Woolhiser, D. A. & McWorther, D. B. (1985) Interrill soil erosion. I. Development of model equations. *Trans. Am. Soc. Agric. Engrs.* **28**, 147-153.
Govers, G. (1985) Selectivity and transport capacity of thin flows in relation with rill erosion. *Catena* 12, 35-49.
Govers, G. (1988) Initiation of motion in overland flow. *Sedimentology* **34**, 1157-1164.
Govers, G. & Rauws, G. (1986) Transporting capacity of overland flow on plane and irregular beds. *Earth Surf. Proc. Landforms* **11**, 515-524.
Kahnbilvardi, R. M., Rogowski, A. S. & Miller, A. C. (1983) Modelling upland erosion. *Wat. Resour. Bull.* **19**, 29-35.
Kalinske, A. A. (1942) Criteria for determining sand transport by surface creep and saltation. *Trans. AGU* **23**.
Kirkby, M. J. (1980) Modelling water erosion processes. In: *Soil Erosion,* (ed. R. P. C. Morgan & M. J. Kirkby), 183-216, Wiley, Chichester, UK.
Komura, G. (1976) Hydraulics of slope erosion by overland flow. *J. Hydraul. Div. ASCE,* **104**, 1573-1586.
Kramer, L. A. & Meyer, L. D. (1969) Small amounts of surface cover reduce soil erosion and runoff velocity. *Trans. Am. Soc. Agric. Engrs.* **12**, 638-641 and 648.
Luque, F. R. & Van Beek, R. (1976) Erosion and sediment transport of bed-load sediment. *J. Hydraul. Res.* **14**, 127-144.
Meyer, L. D. & Monke, J. E. (1965) Mechanics of soil erosion by rainfall and overland flow. *Trans. Am. Soc. Agric. Engrs.* **8**, 572-577.
Moore, I. D. & Burch, G. J. (1986) Sediment transport capacity of sheet and rill flow: application of unit stream power theory. *Wat. Resour. Res.* **22**, 1350-1360.
Morgan, R. P. C. (1980) Field studies of sediment transport by overland flow. *Earth Surf. Processes* **5**, 307-318.
Mossaad, M. E. & Wu, T. H. (1984) A stochastic model for soil erosion. *J. Numer. Analyt. Meth. Geomech.* **8**, 201-224.
Park, S. W., Mitchell, J. K. & Scarborough, J. M. (1982)Soil erosion simulation on small watersheds: a modified ANSWERS model. *Trans. Am. Soc. Agric. Engrs.* 1581-1589.
Parker, G. & Coleman, N. L. (1986) Simple models of sediment-laden flows. *J. Hydraul. Div. ASCE* **112**, 356-375.
Parsons, D. A. (1972) The speeds of sand grains in laminar flow over a smooth bed. In: *Sedimentation, Symp. to Honour Prof. H. A. Einstein, Fort Collins* (ed. H. Shen) 1.1-1.24.
Prasad, S. N. & Singh, V. P. (1982) A hydrodynamic model of sediment transport in rill flow. In: *Recent Developments in the Explanation and Prediction of Erosion and Sediment Yield (Proc. Exeter Symp., July, 1986),* 293-301 IAHS Publ. no. 137.
Rathbun, R. E. & Guy, H. P. (1967) Measurement of hydraulic and sediment transport variables in a small recirculating flume. *Wat. Resour. Res.* **3**, 107-122.
Rauws, G. (1987) The initiation of rills on plane beds of noncohesive sediments. In : *Geomorphological Significance of Rill Development* (ed. R. B. Bryan), Catena Supp. Vol.
Rauws, G. & Govers, G. (1988) Hydraulic and soil mechanical aspects of rill erosion. *J.*

Soil Sci. **39**, 111-124.

Rose, C. W., Williams, J. R. Sander, G. C. & Barry, D. A. (1983) A mathematical model of soil erosion and deposition processes. I. Theory for a plane land element. *Soil Sci. Soc. Am. J.* **47**, 991-995.

Savat, J. (1970) Laboratory experiments on erosion and sedimentation of loess by laminar sheet flow and turbulent rill flow. In *Erosion Agricole en Milieu Tempéré non Mediterranéen*, (ed. H. Vogt), 139-143. Strasbourg.

Savat, J. (1980) Resistance to flow in rough, supercritical sheet flow. *Earth Surf. Processes* **5**, 103-122.

Tödten, H. (1976) A mathematical model to describe surface erosion caused by overland flow. *Z. Geomorphol. Suppl. Bd* **25**, 89-105.

Vanoni, V. A. & Nomicos, G. N. (1959) Resistance properties of sediment-laden streams. *J. Hydraul. Div. ASCE* **85**, 1140-2267.

Wilson, B. N., Barfield, B. J. & Warner, R. G. (1984) A hydrology and sedimentology watershed model. Part II. Sedimentology component. *Trans. Am. Soc. Agric. Engrs* **27**, 1378-1384.

Yalin, M. S. (1963) An expression for bed-load transportation. *J. Hydraul. Div. ASCE* **89**, 221-250.

Yang, C. T. (1972) Unit stream power and sediment transport. *J. Hydraul. Div. ASCE* **98**, 1805-1825.

Yang, C. T. (1973) Incipient motion and sediment transport. *J. Hydraul. Div. ASCE* **99**, 1679-1703.

Erosion, Transport and Deposition Processes (Proceedings of the Jerusalem Workshop, March-April 1987). IAHS Publ. no. 189, 1990.

Evolution of an anthropogenic desert gully system

MARTIN J. HAIGH

Geography Unit, Oxford Polytechnic, Oxford OX3 OBP, UK

Abstract Studies of an unusual quasi-stationary desert gully permit the description of a system attractor which may be inherent, if rarely physically expressed, in many gully systems. Measurements made in a desert gully system (controlled by the seepage of irrigation waters towards a stable base level) suggest a character for the attractor's equilibrium manifold. This is modelled mathematically as an hysteresis or limit cycle in phase space. Overall, the phenomenon is seen to belong to a kind of hierarchical process associated with the Scheidegger "principle of instability/saturation effect". It reflects a dynamic balance between two hierarchical levels inside a single system. The system includes a singularity which is an example of an hierarchical jump caused by the reintegration of a self-assertive holon.

INTRODUCTION

Greene's Canal, on the borders of Pima and Pinal Counties in southern Arizona, is an arroyo. It was created between 1908 and 1910 as a diversion ditch (6.1 m wide by 1.5 m deep) for an agricultural development project. This involved the redirection of flood waters from the Santa Cruz River to a shallow earth-dammed impoundment on the Lower Santa Cruz Plains (Fig. 1). Unfortunately, the dam was destroyed and the project wrecked by severe floods in the years 1914–1915. These floods, called "the worst for generations" (Peirce & Kresnan, 1984), cut the canal into a 3.7 m deep trench, which is twice the depth of the Santa Cruz river at the original point of diversion (Turner *et al.*, 1943; Cooke & Reeves, 1976). At the time of the commencement of this study in 1976, Greene's Canal was 6 m deep and around 80 m wide midway between the diversion and the old reservoir bed.

Greene's Canal is by no means a natural channel. It does not even occupy the topographic low ground. Instead it retains the alignment of its construction which cuts across the contours above the level of the now abandoned natural channel of the Santa Cruz. The arroyo flows north of westward across Quaternary alluvium that slopes gently towards the northeast and the centre of the desert basin. As a consequence, Greene's Canal arroyo has an upslope bank and a downslope bank where the ground falls away from the channel rim.

The cutting of the arroyo trench has steepened the hydraulic gradient above its upslope bank. Existing channels have been rejuvenated and new

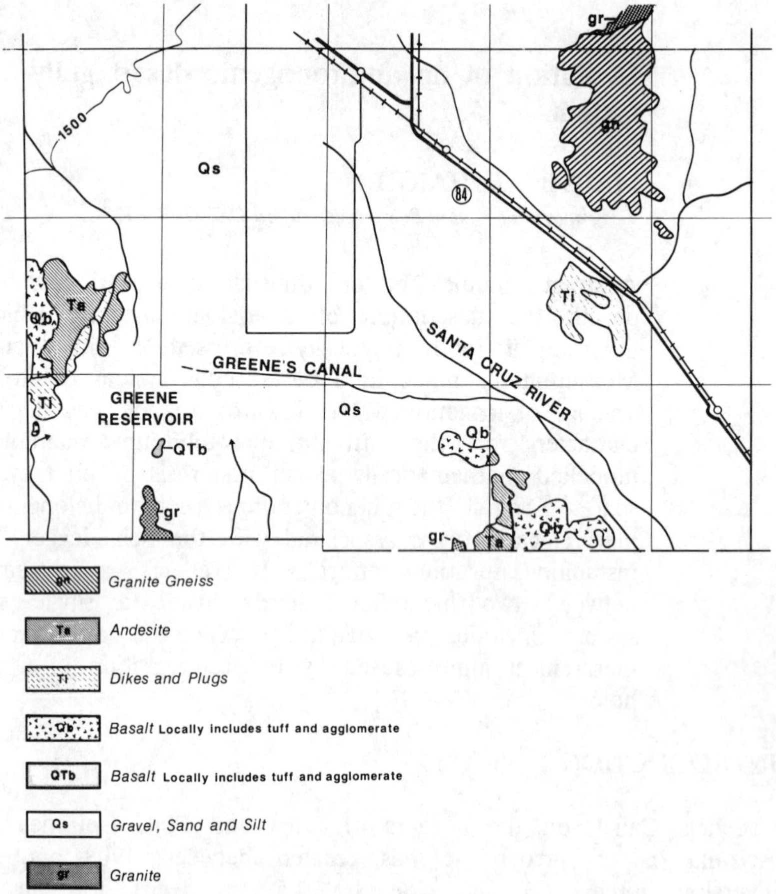

Fig. 1 Greene's Canal, South Arizona, USA.

systems of soil pipes and gully channels have evolved. Comparisons of the USGS topographic survey sheets for 1946 and 1976 show gully displacement of the contours increasing from 500–1000 m to over 1500 m on the arroyo's southwestern flank. By contrast, there has been relatively little gully development on the channel's northeastern, downslope flank. However, some gully channels have evolved here cutting back 20–25 m from the arroyo wall, against the slope of the land.

This study concerns the evolution of one of the gully systems on Greene's Canal's downslope, northeastern flank. This is a very unusual gully in several respects. First, it cuts against the grain of the land. The height of the land at its mouth is almost 200 mm above that at its head cut some 15 m downslope. Secondly, the gully owes relatively little to rainwater. More important is the steady seepage of irrigation waters applied to neighbouring fields. In fact, the lower Santa Cruz Plains are one of the most conspicuous zones of groundwater mining in the USA. This has caused serious and accelerating hydraulic subsidence, amounting to more than 2 m at the point of study, and 5 m at the centre of the desert basin, during the last half century of Greene's Canal's existence (Laney *et al.*, 1978). Subsidence has opened up a network of fissures around the margins of the basin and it has increased the slope across which Greene's Canal flows. Coincidentally, it has back tilted the channel of the study gully, albeit very slightly.

Fortunately, these singularities make this tributary gully of Greene's Canal arroyo of very special research significance. Firstly, it demonstrates the pattern of evolution of a gully whose behaviour is entirely controlled by groundwater seepage. This gully has virtually no catchment for surface water beyond its own walls. Secondly, and in contrast to most desert gullies, this gully can be examined as a discrete system. From the emplacement of erosion monitoring equipment in September 1976 and until its destruction during the floods of October 1983, there was insufficient rainfall to cause incision in the main arroyo. The base level of the gully remained constant during 6.6 years of data collection. Meanwhile, gully evolution continued with the help of regular inputs of irrigation water seeping from nearby fields. Thirdly, therefore, this gully is a representative of a particular breed of desert gully, which is liable to evolve wherever unlined canals run close to irrigated fields.

The behaviour of this gully system also has special theoretical significance. This is because its pattern of evolution includes a stationary cyclic element or torus of the type described by Scheidegger (1983) as the geomorphological "instability principle/saturation effect" and by Bennett & Chorley (1978) as stability in the sense of Liapunov.

The cyclic element includes a long period of slow progressive evolution and a short-lived phase of hysteretic reversion, which suggests that it might be modelled through the medium of catastrophe theory. However, more important is the fact that this pattern is a rare demonstration of how an active gully system may exist in "dynamic equilibrium" with its environment. It is argued that this stability is possible because the system is operating close to an intrinsic system attractor. It is the character of this attractor which this paper seeks to identify.

THE TEST GULLY

The gully examined for this project is typical of those on the downslope bank of Greene's Canal. It is 15 m long from mouth to channel head cut. Its average slope is 6% if the most recent deep head cut is ignored and 17% if it is included. The channel long-profile is broken by two head cuts, a degraded 100 mm step just 2.2 m from its final head cut, and a deep active head cut near the mouth of the gully (Fig. 2).

Gully morphometry has been recorded by means of a 150 mm slope pantometer. Survey included the original long profile and cross profiles recorded at 2 m intervals between mouth and head cut (Fig. 3). One of the seven cross profiles coincided with a minor confluence and this was abbreviated at the mid point of the channel.

Erosion pins are established at regular intervals along each cross profile. A record of changes in the exposures of these erosion pins was kept for 6.7 years between September 1976 and April 1983. The results of these

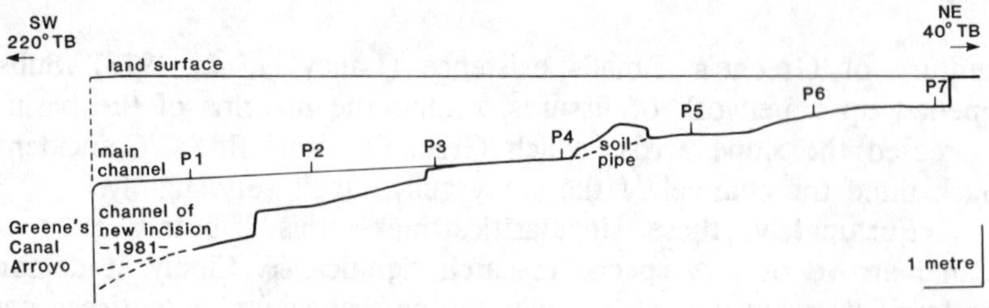

Fig. 2 Long profile of test gully showing locations of erosion pin gross profiles.

measurements of total ground loss in millimetres are recorded for individual sites on Fig. 3.

A detailed discussion of the empirical results has been published by Haigh & Rydout (1987). In summary, what happens is that the gully banks tend to retreat rapidly and parallel to themselves. In the process, a small, compacted slope foot segment and a wide flat depositional basin are created. This process is interrupted periodically, when the roof of a subterranean soil pipe breaches the gully floor creating a narrow, vertical sided slot in the gully basin. The channel at the base of this trench quickly becomes choked by the debris released from its own walls. As these walls retreat, the trench becomes wider and more shallow (Fig. 4).

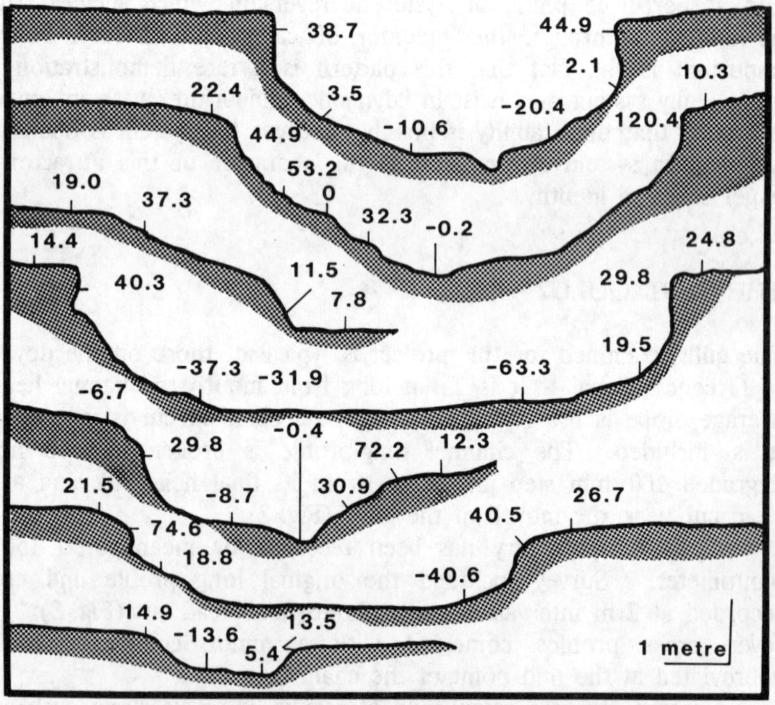

Fig. 3 Cross profiles of test gully showing data collected at each erosion pin (ground retreat in mm: total for 6.6 years of study).

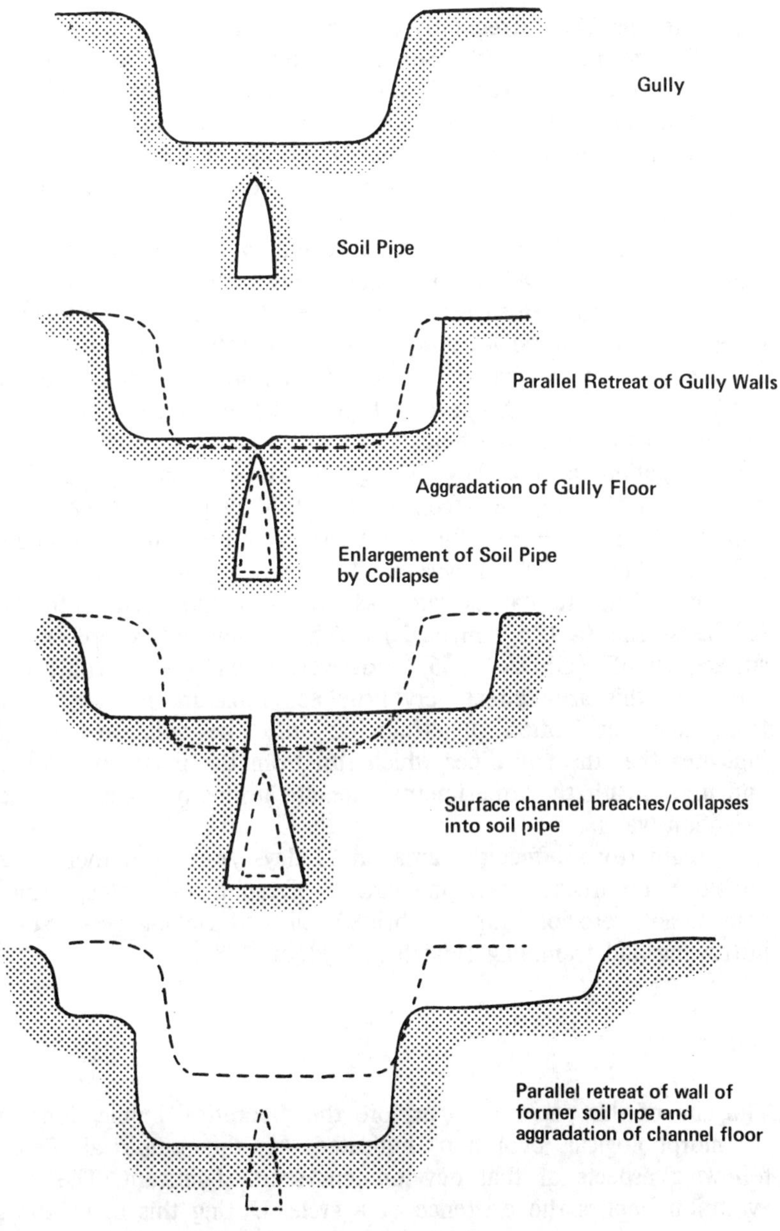

Fig. 4 Collapse/fill cycle of test gully basin.

SITE CHARACTERISTICS AND PROCESSES

The test gully is cut into soils which are Typic Torrifluvents, perhaps mixed with Aridic Cumulic Haplustolls. They are fine textured (90% passing the 63 μm sieve), calcareous (pH 7.9–8.4), slowly permeable (5–15 mm h^{-1}) loams of low density (1.35–1.80 g cm^{-3}) and high erodibility (K = 0.32–0.37). However, the soils include thin clayey layers of higher density and lower erodibility dispersed irregularly through the profile.

The ground beneath the gully's floor is cut by a network of large soil pipes. Where they break through into the walls of the main arroyo, these triangular pipes may be 2 m high and 0.30–0.5 m wide at their base. Collapse hollows are common even 5 and 10 m beyond the head cuts of the surface gully fringe to the arroyo, and subsidence ruts can be discovered in the arable fields beyond.

The test gully channel was unvegetated throughout the study except for a brief period in 1983 when a 30% grass cover developed on its floor and lower north wall. However, the channel is cut into a platform dominated by creosote bush (*Larrea tridenta*). At the mouth of the gully, the arroyo floor is colonized by a mature scrub of mesquite (*Prosopis sp.*) and saltbush (*Atriplex polycarpa*). Some 5 m beyond the gully head cut are pasture grasses on the margins of arable fields.

Irrigation is the key to agriculture in this region. Flood irrigation methods involve anything from 0.37 to 0.74 m year^{-1} of water application. By contrast, rainfall during the 6.7 years of measurement totalled only 1.9 m (277 mm year^{-1}), which was a little above average (248 mm year^{-1}). The erosivity of the rainfall is rated as low (R = 75), despite the fact that 62% falls as storms (> 12.5 mm day^{-1}) which are arguably severe enough to activate surface runoff (USDA 1966). However, Cable (1977) confirms that rainfall events in this area effect very little soil moisture recharge and the wetting front does not often penetrate far into the soil. In sum, the evidence indicates that the soil pipes which run from the irrigated fields to the arroyo and underneath the experimental site are primarily caused by the drainage of irrigation waters.

Light frosts affect the area on 11 days year^{-1}, but there is very little soil moisture to freeze even in winter. Despite this, steep soil faces suffer considerable erosion due to shrink/swell and slaking processes, and also to burrowing and trampling (Haigh & Rydout, 1987).

ANALYSIS

The aim of this paper is to explore the theoretical implications of the pattern of morphological evolution indicated by the empirical field study. The following aspects of that development seem significant. The pattern of gully evolution implies the existence of a cycle. During this cycle, the gully exhibits two mutually exclusive models of behaviour. Mode 1 is where morphogenesis is led by gully side wall retreat and aggradation of the depositional gully basin floor. Mode 2 is where morphogenesis is led by the exposure, mainly if not exclusively by collapse, of the soil pipe growing within the deposited sediments of the gully basin. Mode 2 morphogenesis operates through a much shorter time scale then mode 1. However, both can be occurring concurrently at different points within the same gully basin. Furthermore, the possibility exists for the two processes to hold each other in check as a kind of limit cycle. Mapped in phase space, the relationship can be represented as a torus (Kaneko, 1986).

MATHEMATICAL MODEL

The two process systems involved must first be considered separately. First, is the normal, at least in terms of duration, pattern of gully evolution.

Under "normal" conditions, the elevation of the gully floor at the instant of surface collapse (R = 0) depends upon the depth at which the soil pipe channel floor forms. This seems to be close to the upper surface of the water table created by the seeping irrigation waters. Measurements made in pipe collapse hollows indicate that this ranges from around 1.5 m close to the arroyo wall to around 0.5 m at the outer fringe of the gullied area.

Gully depth begins as the difference between base level (R = 0) and the height of the original ground surface (K). However, as time passes, gully depth ($K - R$) declines as the channel floor is aggraded by the transfer of sediments released from the pipe/gully side walls. The morphological consequence of this is that the side walls shrink and retreat as the gully basin expands. As the gully becomes more shallow and wider towards the pre-gully ground surface (K), the rate of aggradation (x) declines asymptotically.

$$\frac{\mathrm{d}\,(K - R)}{\mathrm{d}t} = (K - R)^{x} \qquad (1)$$

where: x is the exponent representing the rate of channel aggradation;
K is the elevation of the original ground surface;
R is the initial depth of the gully channel; and
t is time.

The rate of aggradation (x) is a function of the rate of sediment release from the channel banks (ca. 8 mm year^{-1}), scaled by changes in packing density (the ratio between the bulk densities of the channel bank and channel floor sediments), and controlled by the progressively declining ratio between gully floor and gully side.

However, underground, the soil pipe is a water scoured channel. On exposure, this channel quickly becomes buried due to the massive release of sediments from the gully walls. As it becomes buried, it requires more and more energy to generate surface water flows. The frequency of flows and the volume of sediments cleared declines. Scour decreases further as wash due to rainwater becomes spread across the floor of an increasingly large depositional basin with progressively reduced local relief. Eventually, surface aggradation in the gully comes to depend only on the rate of side wall retreat counterbalanced by the rate of scour due to surface wash processes.

Adding these complications, transforms the pattern of "normal" morphogenesis from a negative exponential curve to a sigmoid curve. The equation, then, may be rewritten thus:

$$A = (K - R)^{x} - S - P \qquad (2)$$

where: A is channel shallowing (mm year^{-1})

$(K - R)$ is gully depth (mm);
x is an empirical exponent with a value of 0.26;
S is channel deepening due to scour, which in this environment, witness the crest segment retreat rates, is at least 2.7 mm $year^{-1}$;
R is current elevation of the gully floor (mm);
K is the elevation of the ground surface (mm); and
P represents scour associated with soil pipe flows (mm).

Naturally, the values of x, P, and S may be expected to vary with soil type, climate, irrigation levels, and variations in the initial width of the gully channel at the start of a new cycle. The values cited here are calculated from the results of this single case study. This includes few good measurements in the soil pipe scour zone, so this term is not quantified. However, the influence of P appears to be restricted to the zone below R = 150 mm. The field data, however, do permit the quantification of the balance of the process. So, solving Equation (2), it can be seen that at 2 m depth, the maximum rate of aggradation would be 7.2 mm minus S = 2.7, which leaves a net channel aggradation of 4.5 mm $year^{-1}$. At a depth of 200 mm the rate of aggradation would be 4.0 mm minus 2.7, leaving 1.3 mm $year^{-1}$ net aggradation (cf. Fig. 4).

However, the numbers gleaned from this case study are much less important than the pattern they suggest. It now becomes possible to represent this gully collapse and fill cycle as a graph (Fig. 5), and the problem reduces to finding a term for the collapse phase of the system. This is the period during which the surface gully collapses into the soil pipe evolving within its own deposits.

Several aspects of the process affect any mathematical model. First, once the initial stages of channel infilling bury the groundwaters, the two systems operate completely independently one of the other. The evolution of the soil pipe has no impact on the surface until the commencement of collapse. Second, nothing is known about the character of the expansion of the soil pipe prior to its exposure at the ground surface. However, because this pipe is developing within a relatively stable energy stream, the throughput of irrigation waters, it might be suggested that for a stable pipe/gully system, the phase change might be identified as a threshold depth of deposited sediments (T).

Naturally, any such threshold is only identifiable as a statistical mode

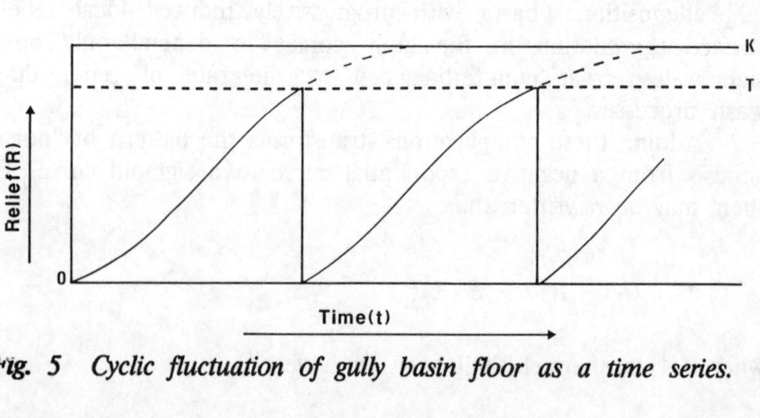

Fig. 5 Cyclic fluctuation of gully basin floor as a time series.

or mean. However, there are few head cuts in this gully system or its neighbours which are smaller than 0.5 m and very few which are deeper than 2.5 m. Most fresh collapse hollows lie in the range 0.5–1.5 m. Indeed, the fact that collapse hollows exist apart from the major gully systems indicates that ground surface and soil pipe can co-exist through quite a large range of intervening soil depths.

The second important aspect of the collapse part of the cycle is that it occurs very abruptly in the terms of the normal operating time scales of the system. The elimination of the work accomplished by the "normal" operations of the surface system occurs within days rather than the decades of the full cycle.

In mathematical terms, the operations of this unrecordable subterranean control of the gully channel can be conceived as the equivalent of a modulus. The system is regulated by modulo T, which may be approximated in a stable system as a critical depth of burial of the water percoline at $R = 0$.

$$\frac{d(K - R)}{dt} = (K - R)^{x} - S - P : *T \text{ (modulo } T) \tag{3}$$

where:

- T is the threshold depth of sediment infilling where collapse into the underlying soil pipe usually occurs (500 mm < 1500 mm);
- R is the height of the gully floor above its base level (R_o = 0 mm);
- K is the altitude of the original surface above R_o (mm);
- x is the exponent for gully infilling;
- S is channel deepening due to scour (mm); and
- P represents scour associated with soil pipe flows (mm).

CATASTROPHE THEORY

The problem with the foregoing analysis is, of course, the fact that the phenomenon of surface collapse is not really controlled by a modulus. Between depths of burial of 0.5 to 1.5 m, surface gully and soil pipe can co-exist independently, or because of local circumstances, they may interact through collapse. The system can exist quite easily in either of two distinct conditions through a large part of its range of operations. It can, abruptly, move from one state to the other unpredictably at any moment during the period when these two possibilities exist. In mathematical terms, the proper description for this condition is a catastrophe (Saunders, 1980).

Catastrophe theory is NOT a theory. It is merely a mathematical language for describing some types of system which contain discontinuities or abrupt changes of a particular character. It was developed by Thom (1972) as an alternative to calculus, which is a language for examining continuous or smooth changes. Catastrophe theory is a branch of the mathematical science of topology which deals with phenomena both numerically and as geometrical forms. The catastrophes which it describes are not necessarily "disasters". They are merely species of phenomena which jump abruptly from one mode of behaviour to another because of the co-existence of two basins of attraction in the system.

Since its creation, catastrophe theory has been both in and back out of fashion. It has been both used and abused (Zahler & Sussman, 1977; Thompson, 1982). It has been applied to geomorphology and geology (Henley, 1976; Wilson, 1981). The truth is that the language of catastrophes has major limitations. First, this language applies only to phenomena which can be described by a potential function - where system behaviour is a consequence of something being maximized or minimized across a mappable gradient. Second, its models only hold locally, in the vicinity of the critical state. Finally, the language is entirely descriptive and not at all predictive (Wilson, 1984).

There are several major classes of catastrophe. Each can be defined in terms of a number of control functions and behaviour factors. The simplest catastrophe has a single control factor and a single behaviour function which describes a folded graph with a single maximum and a single minimum (Woodcock & Poston, 1974). The cusp catastrophe includes the fold catastrophe and an extra control factor or dimension along which the degree of the fold varies. The result can be pictured in three dimensions and its behaviour function has two minima and one maximum. In effect, the cusp catastrophe organizes two one parameter families of fold.

Higher order catastrophes have larger numbers of control and behavioural dimensions. Their full geometry is multidimensional and hence rather difficult to portray and conceive (Woodcock & Post 1974). To date, there have been few attempts to apply catastrophe models above the level of the simple cusp.

A CATASTROPHE MODEL FOR THE DESERT GULLY SYSTEM

Examined as the graph of Fig. 5, an individual cycle of system behaviour is explained in terms of one control dimension, gully depth, and one behaviour dimension, aggradation/collapse. So the graph portrays a simple fold catastrophe which is repeated as a regular cycle bounded by threshold (T) and base level ($R_o = 0$).

The location of the fold is determined by the soil strength variable which defines the location of the collapse threshold ($K - R = 500 < 1500$ mm approximately) and hence both the point of bifurcation in the system and the orbit of the torus. This control of the bifurcation may be added to the model in terms of a second control dimension. This variable is $(K - R_o)$ which varies with the depth of the soil pipe and hence controls the expression of the catastrophe as well as the pattern of aggradation. The behavioural dimension (A = solution of Equation (3)) can then be mapped across the two control dimensions: $(K - R_o)$ and time (t). This additional control organizes the family of fold catastrophes as a cusp catastrophe.

This surface requires an equation of the form

$$V\,(A,\ (K - R_o),\ t) = A^4 + A^2 \cdot (K - R_o) + A \cdot t \qquad (4)$$

where V is the system potential (Thompson, 1982).

DISCUSSION

In geomorphological theory, Scheidegger (1983) illustrates his principle of instability by reference to the behaviour of river meanders. Every river meander tends to divert its flow of water to the outside of the meander bend and so increase its curvature. River meanders belong to that large class of landforms which do not tend to a steady state (Bull, 1975). However, river meanders are also subsystems (holons) within the larger system of the river channel (Haigh, 1987). This has work to do as is witnessed by the way that rivers' long profiles hold to grade (Bull, 1979). The self-assertion of the meander holons reduces the efficiency of the operations of the larger system. Eventually, that larger system is forced to re-integrate its member holons. A surge of energy in the system causes the re-establishment of the original energy line (channel way), the development of meander cutoffs and so the restructuring of the meander holon.

The behaviour of the Greene's Canal tributary gully system is not very different in character. The gully is part of a larger system, the drainage of irrigation waters from arable fields to neighbouring arroyo. The gully emerges because the interactions between the seeping waters and the desert soils lead to the development of soil pipes. These pipes expand upwards by roof collapse until they become exposed at the soil surface. At this point, the gully/soil pipe may be performing its role as a sub-system (holon) efficiently. However, sub-aerial processes tend to the elimination of surface gullies. The seepage water flow-away becomes buried and the gully begins to heal. However, the original energy line still lies buried in the soil, so eventually the soil pipes reform and the whole cycle goes through another loop.

The significance of the relationship is, however, best understood in the terms of hierarchy theory. The gully can be examined as a holon within the larger system of relationships linking the irrigated fields to the nearby arroyo. Normally, morphogenesis in the gully continues independently of higher level control. However, these processes decrease the efficiency of operations at the higher level. Eventually, this higher level reasserts its control over the sub-system. It is this hierarchical restructuring which is the system catastrophe. It is this relationship between two hierarchical levels in the system which the mathematical models really describe.

The relationship, then, is an example of interaction between two hierarchical levels in a single system. The catastrophe is an example of hierarchical restructuring (Haigh, 1987; Platt, 1970). The cusp model control factors represent the antagonistic agenda of the two holonic levels involved.

Similarly, the linear model of sigmoid growth regulated by a modulus, is a first cousin to the predator-prey equations which describe the interactions between two hierarchical levels in any ecosystem (Thompson, 1982). The predator-prey relationship with its clear pattern of stable oscillation is widely recognized as a significant system attractor in ecological systems. The status of the gully system attractor demonstrated here is rather less certain. Certainly, rill systems often show a similar dynamic balance between incision due to flowing water and healing due to other sub-aerial processes. Blong *et al.* (1982) have argued that the role of side wall processes is frequently understated in studies of desert gullies. Furthermore, the fact that some gully systems can remain stable over many centuries is now well established. It may well be that the character of the "dynamic equilibrium" described at Greene's Canal may not be so unusual, and that the system attractor here described is merely underrecognized.

CONCLUSION

The results of a 6.6 year period of erosion pin measurement in a desert gully indicate that for most of the time morphogenesis is controlled by the retreat of the gully side walls and deposition in the gully basin. However, this is interrupted occasionally by the collapse of the gully floor into underlying soil pipes. These pipes are created by the seepage of irrigation waters between arable fields and the Greene's Canal arroyo to which the gully is tributary. The end product of this control is a gully system which, to all intents and purposes, is held in a stationary limit cycle regulated by a catastrophe (hierarchical jump). This cycle may provide an attractor for many arid gully systems and may help explain some of the morphological complexity of relatively stable gullied landscapes.

Acknowledgements Figures 1–4 are reproduced, by kind permission of J. Wiley and Sons Limited, from *"International Geomorphology 1986, Part 2"*, edited by V. Gardiner and published in 1987. The author would like to thank Gary B. Rydout for his continuing support of the Greene's Canal project.

REFERENCES

Bennet, R. J. & Chorley, R. J. (1978) *Environmental Systems: Philosophy, Analysis and Control.* Methuen, London.

Blong, R. J., Graham, P. O. & Veness, J. A. (1982) The role of sidewall processes in gully development, some N.S.W. examples. *Earth Surf. Proc. Landforms* 7(6), 381-386.

Bull, W. B. (1975) Allometric change of landforms. *Geol. Soc. Am. Bull.* **86**, 1489-1498.

Bull, W. B. (1979) The threshold of critical power in streams. *Geol. Soc. Am. Bull.* **90**, 453-464.

Cable, D. R. (1977) Soil water changes in creosote bush and bursage during a dry period in southern Arizona. *Arizona Acad. Sci. J.* **12**(1), 15-20.

Cooke, R. U. & Reeves, R. W. (1976) *Arroyos and Environmental Change in The American Southwest.* Oxford Research Studies in Geography, Clarendon Press, Oxford, UK.

Graf, W. L. (1979) Catastrophe theory as a model for change in fluvial systems. In: *Adjustments of the Fluvial System, Binghamton Geomorphology Symp.* **10**, (ed. D. D. Rhodes & G. P. Williams), 13-32, Kendall Hunt, Dubuque, Iowa, USA.

Haigh, M. J. (1987) The holon, hierarchy theory and landscape research. *Catena, Suppl.* **10**, 181-192.

Haigh, M. J. & Rydout, G. B. (1987) Erosion pin measurements in a desert gully, Greene's Canal, Arizona, in: *International Geomorphology 1986* (ed. by V. Gardiner), 419-436, Wiley, Chichester, UK.

Henley, S. (1976) Catastrophe theory models in geology. *Math. Geol.* **8**(6), 649-662.

Kaneko, K. (1986) *Collapse of Tori and the Genesis of Chaos in Dissipative Systems.* World Scientific, Singapore.

Laney, R. L., Raymond, R. H., & Winikka, C. C. (1978) Maps showing water-level declines, land subsidence and earth-fissures in south-central Arizona. *USGS Wat. Resour. Invest.* 78-83,

Open File Report.
Saunders, P. T. (1980) *An Introduction to Catastrophe Theory.* Cambridge University Press, Cambridge, UK.
Peirce, H. W. & Kresnan, P. L. (1984) The floods of October 1983. *Arizona Bureau of Geology and Mineral Technology, Field Notes* **14**(2), 1-7.
Platt, J. (1970) Hierarchical restructuring. *General Systems* **15**, 49-54.
Scheidegger, A. E. (1983) Instability principle in geomorphic equilibrium. *Z. Geomorphol. N.F.* **27**(1), 1-19.
Thom, R. (1972) *Structural Stability and Morphogenesis.* (trans: Fowler, D. (1975)) Benjamin, Reading (in Thompson, 1982 op cit).
Thompson, J. M. T. (1982) *Instabilities and Catastrophes in Science and Engineering.* Wiley, Chichester, UK.
Thornes, J. (1983) Evolutionary geomorphology. *Geography* **68**(3), 225-235.
Turner, S. G. *et al.*, (1943) Groundwater resources of the Santa Cruz Basin, Arizona. *USGS Open File Report.*
USDA (1966) Procedures for determining rates of land damage and depreciation and volume of sediment production by gully erosion. *US Department of Agriculture, Soil Conservation Service, Technical Release (Geology)* **32**.
Wilson, A. G. (1981) *Geography and the Environment, Systems Analytical Methods.* J. Wiley, Chichester, UK.
Wilson, A. G. (1984) Catastrophe theory. *Times Higher Education Supplement* 25.5.84, 15.
Woodcock, A. E. R. & Poston, T. (1974) *A Geometrical Study of the Elementary Catastrophes.* Lecture Notes in Mathematics 373, Springer-Verlag, Berlin.
Zahler, R. S. & Suusman, H. J. (1977) Claims and accomplishments of applied catastrophe theory. *Nature* **269**, 759-763.
Zeeman, E. C. (1976) Catastrophe theory. *Sci. Am.* **234**, 65-83.

Erosion, Transport and Deposition Processes (Proceedings of the Jerusalem Workshop, March-April 1987). IAHS Publ. no. 189, 1990.

The relationship between sediment delivery ratio and stream order: a Romanian case study

IONITA ICHIM

Research Station "Stejarul", str. Alexandru cel Bun, 6, Piatra Neamt, 5600, Romania

Abstract An investigation of sediment delivery ratios in three regions of Romania (Moldavian Tableland, Subcarpathians and Flysch Mountains) shows a inverse relationship between sediment delivery ratio and drainage basin order (Strahler's system). However, these relationships vary in their precise form. There are two main controlling factors, namely, rock erodibility and runoff regime. When the rocks are easily eroded and the small headwater catchments evidence runoff of torrential character, the sediment delivery ratio decreases markedly with an increase in river network order. This is the case in the Moldavian Tableland, where the foot hills are a zone of active colluviation. When the rocks are easily eroded, but the runoff is greater, high sediment delivery ratios exist. This is the case in the Subcarpathians. When the rocks are resistant to erosion and the runoff is also high, the relationship between sediment delivery ratio and drainage basin order exhibits a more moderate slope, as shown by the Flysch Mountains.

INTRODUCTION

An investigation of the relationship between sediment delivery ratio (SDR) and drainage basin order (Strahler's system) has been undertaken for three areas of Romania, namely, the Moldavian Tableland, the Subcarpathians and the Flysch Mountains. The last two areas, are represented by the basin of the Putna River in the Vrancea region (Fig. 1). The slope erosion was designated the *effective erosion* (E_v t km^{-2} year^{-1}) and the *specific rate of sediment transfer* (E_s t km^{-2} year^{-1}) was estimated from measurements of sediment load at river cross sections. E_s therefore also represents the *sediment yield.* In the areas in which slope erosion is primarily the result of surface erosion, the distinction between effective erosion and sediment yield is clear. On the contrary, in areas in which sediment production from the slopes is mainly the result of mass movements (landslides, mud flows, creep) only loosely connected with fluvial processes, the distinction is less clear. In these cases, it is better to think in terms of *gross erosion* which is defined as the sum of inter-rill, rill, gully and channel erosion in small catchments. The sediment delivery ratio represents the ratio between the effective erosion or the gross erosion and the sediment yield.

In the cases studied, *effective erosion* was employed for the Moldavian Tableland, where the main sediment sources from the slopes are associated with surface erosion. For the Flysch Mountains and Subcarpathians areas, *gross erosion*, representing the sediment yields for basins of order I and II respectively (Ichim & Radoane, 1986), was used. These basins are smaller than 1 km^2.

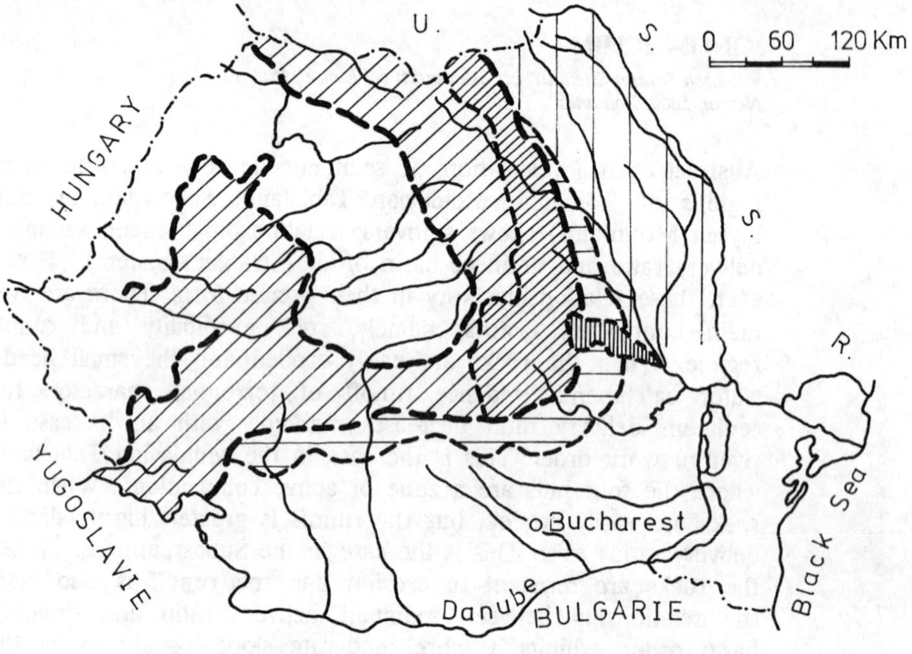

Fig. 1 Location of the study areas within Romania.

MOLDAVIAN TABLELAND

This area, situated in the Eastern part of Romania, with an area of about 25 000 km^2, is characterized by soft sedimentary strata, with clays and marls alternating with sands, gently dipping in a southeastern direction (5–8%). The average annual precipitation is 450–600 mm and the greatest part of the land is used for agriculture or pastures. Severe or very severe erosion exists on about 50% of the land surface of the whole region, with most of the erosion resulting from heavy storms of over 20 mm.

Plot experiments undertaken for over 30 years (Popa, 1977; Baloiu & Giurma, 1979; Ionita & Ouatu, 1985) indicate that on the fields under cereals, the *effective erosion* varies between 225 t km^{-2} $year^{-1}$ and 5550 t km^{-2} $year^{-1}$, with a general average of about 2700 t km^{-2} $year^{-1}$. If the gully sediment yields are also considered, the data obtained from over 20 years of field experiments indicate that this is on average about 3150 km^{-2} $year^{-1}$

(Motoc *et al.*, 1979). On the Moldavian Tableland the average effective erosion is therefore at least 2250–3000 t km^{-2} $year^{-1}$.

Date relating to measurements undertaken for periods of over 20 years indicate that from basins of order IV and greater, sediment yields vary between 14 and 694 t km^{-2} $year^{-1}$. It is, however, important to note a special situation which strongly influences the sediment delivery ratio from the Moldavian Tableland. This relates to the presence of about 800 small reservoirs, some up to 400–600 years old, which retain some of the sediment transported by the rivers. The amount of sediment deposited in some of the reservoirs has been evaluated in order to take into account the trapping of sediment in the reservoirs. The measurement of sediment yield from the drainage basins have therefore been corrected in order to estimate the natural sediment delivery ratios which have been related to the Strahler order data. (Table 1, Fig. 2). These data demonstrate two important trends. Firstly, the sediment delivery ratios decrease very rapidly up to the level of the basins of order IV. This can be accounted for by the importance of local storms during much of the year. Only during the spring floods does sediment transport extend over all the drainage basins larger than those of order IV.

Table 1 *The variation of sediment delivery ratio (%) with channel network order in the three study areas*

Region	*Channel network order (Strahler's system)*						
	I	*II*	*III*	*IV*	*V*	*VI*	*VII*
Flysch Mountains	*100.0*	*65.2*	*42.2*	*33.2*	*26.1*	*20.0*	-
Subcarpathians	-	*100.0*	*80.9*	*61.6*	*45.6*	*30.0*	*25.0*
Moldavian Tableland	*100.0*	*49.5*	*34.6*	*19.0*	*12.0*	*5.5*	*3.5*

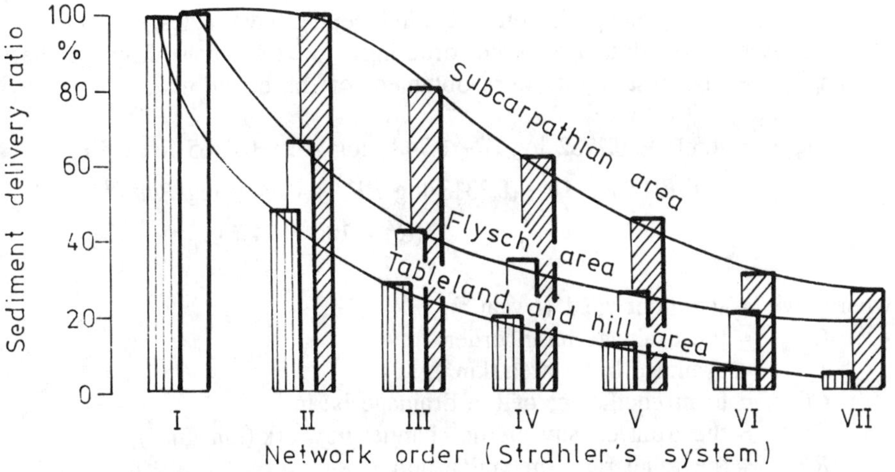

Fig. 2 *Sediment delivery ratio related to channel network order.*

Secondly, under the present morphoclimatic conditions of the Moldavian Tableland the very great difference between the effective erosion rate and the sediment yield of the drainage basins, is a reflection of the powerful colluviation processes and the active deposition of sediment at the base of the slopes.

THE SUBCARPATHIANS

Within the territory of Romania, between the Carpathians and the extra-Carpathian hilly region, there is an area of geomorphological and geological transition, called the *Subcarpathians.* This is developed on neogene molasse deposits which have been folded and which show a marked tendency for mass movements (landslides, mud flows, earth flows etc). These phenomena considerably increase the slope erosion rate. Up to the basins of order III, any differentiation between the contributions from the slopes and from the headwater river channels to the sediment yield is very difficult. This is due to the fact that a large proportion of the mass movements (especially, landslides, earth flows, mud flows, creep, etc) are produced as a direct response to gully erosion. For this reason, the sediment yield of the IInd order basins has been considered to represent the *gross erosion*.

The Subcarpathians have an average altitude of 550 m and a maximum altitude in excess of 1000 m. The annual precipitation lies between 550 and 800 mm, with a high frequency of storms. The land is mainly used for grazing, forests and agriculture.

Sediment delivery ratios were evaluated for a sample of basins from the Putna River drainage basin in Vrancea. Measurements of sediment yield have been made for more than 20 years. The results show that the gross erosion frequently attains levels of 4000–5000 $m^3\ km^{-2}\ year^{-1}$ (Gaspar *et al.*, 1982). This is one of the highest erosion rates in the Carpathian region. Several prediction equations have been developed by Ichim & Radoane (1986), based on multivariate statistical analyses of the relationships between sediment yields and different controlling factors (28 independent variables, including the Strahler system of drainage basin ordering) for 63 small drainage basins. Among these, the best results were obtained for the equation:

$$\log Sy = 4.502 - 0.1782 \log \Omega + 0.7486 \log A^{a} + 0.0365 \log Cf + 0.1042 \log Dt + 0.3318 \log RR^{b} + 0.5439 \log Pmm^{d}$$

$$(R^2 \times 100 = 74\%)$$

where: *Sy* = sediment yield ($t\ year^{-1}$);
Ω = the drainage basin order;
A = drainage basin area (km^2);
Cf = form coefficient of the drainage basin;
Dt = the total density of the channel network ($km\ km^{-2}$);
RR = the relief ratio ($m\ km^{-1}$); and
Pmm = mean annual precipitation (mm).

Significance levels:

a = very high significance level (99.9%); *b* = high significance level (99.5%); *d* = low significance level (99%).

This equation was used to determine the sediment yield of small drainage basins in the Vrancea Subcarpathians and values close to those registered by field measurements were obtained.

To select the best function to represent the relationship between sediment yield and channel network order we tested 20 functions from which a computer program selected the optimum equation of the form:

$$Sy = \left[-17\ 918.84 + 25\ 062.81\ \Omega\ (-3563.21\Omega^2)\right]/\Omega^2$$

$$(n = 63;\ r = 0.557)$$

in which Sy = sediment yield (t year^{-1}) and Ω = order of the drainage basin. Using this relationship we obtained for the drainage basins of order II (A < 1 km^2) Sy = 4538 t km^{-2} year^{-1}. This was taken as an average value of gross erosion. It is comparable with the values obtained in some experimental basins by Gaspar *et al.* (1982) and by means of the equation proposed by Ichim & Radoane (1986). This value enabled us to calculate the sediment delivery ratios shown in Table 1 and Fig. 2. To check the results obtained for the different network orders we used the measured data for some river sections on the Putna river.

When compared with the Moldavian Tableland and the Flysch Mountains, the sediment delivery ratios in the Subcarpathians are the greatest. This can be accounted for by contrasts in the magnitude and frequency characteristics of the discharge regime (the flow volume in the Subcarpathians is generally greater than that in the Moldavian Tableland), by differences in the degree of erodibility of the rocks (greater in the Subcarpathians than in the Flysch Mountains) and by the increased incidence of landslides and mud flows and their role in the direct transfer of sediment into river channels.

THE FLYSCH MOUNTAINS

In the Septentrional and Eastern Carpathians, the Flysch represents one of the most extensive lithological complexes. Within Romania's territory it extends to a maximum of almost 100 km. One of its major characteristics is the alternate layers of sandstone, marls, clay schist and sometimes conglomerates which are strongly folded.

The maximum altitude is 2508 m (Omul peak in the Bucegi Mountains), but except for this mountain, the area lies below 2000 m and has an average altitude of 300–500 m. Unlike the Subcarpathians, mass movement processes play a more limited role in the transfer of material from the slopes into the river channels. With a few exceptions, almost the whole mountain area is covered in mixed woodland. The mean annual precipitation lies between 600–1200 mm year^{-1} with a monthly maximum in June, and a lack of severe storms. Under these conditions the effective erosion rates and the sediment yields are much smaller than in the Subcarpathians.

As in the case of the Subcarpathians, the sediment yields of small drainage basins (order I, Strahler's system) have been taken to represent the gross erosion.

Sediment yield was evaluated using an equation of the form:

$$\log Sy = 7.985 + 0.8138 \log A - 0.304 \log Rb + 0.1486 \log Dt$$
$$- 0.1547 \log RR + 0.089 \log Af - 1.571 \log Pmm$$
$$(R^2 \times 100 = 92\%)$$

where Rb = bifurcation ratio; and Af = afforested coefficient (%) established by Ichim & Radoane (1986) for small drainage basins in the Flysch Mountains. To examine the relationship between sediment yield and the order of the drainage basin, the same procedure as used for the Subcarpathians area was followed. This produced the relationship:

$$Sy = \left[-4493.45 + 5826.65\ \Omega\ (-671.06\ \Omega^2)\right] / \Omega^2 \quad (n = 36;\ r = 0.606)$$

The gross erosion was estimated by this equation to lie between 1000–1500 t km^{-2} year^{-1}, with an average of 1255 t km^{-2} year^{-1} for the basins of order I. A comparison of these data with the results of the experimental investigations of small drainage basins and experimental plots (Arghiriade *et al.,* 1960; Ichim *et al.,* 1979; Radoane, 1981; Gasper *et al.,* 1982) show that the results are reasonable. On this basis, the sediment yield of the Putna river basin (Vrancea) was determined for network orders up to order IV. Measurements from several river cross-sections were also available. The sediment delivery ratios listed in Table 1 and Fig. 2 were thereby calculated.

The sediment delivery ratios of the Flysch areas are greater than those of the Moldavian Tableland, but smaller than in the Subcarpathians. The increased incidence of more resistant rocks causes the material transferred from the slopes into the river channels to be coarser, and therefore the rates of transfer to be slower. The suspended sediment loads are therefore less than those transported by the Subcarpathians rivers where clays and marls, the more highly erodible rocks, dominate. This is despite the fact that the runoff is greater than in the Moldavian Tableland and Subcarpathians. It must also be noted that the mass movement processes, which can be important in the transfer of materials from the slopes into the river-channels, are not as significant as those in the Subcarpathians.

CONCLUSIONS

The variation of the sediment delivery ratios in the three areas investigated falls within the limits of the general tendencies identified for other areas of the world. This is further indicated by the relationships between sediment yield and drainage basin shown in Fig. 3. The marked differences between the

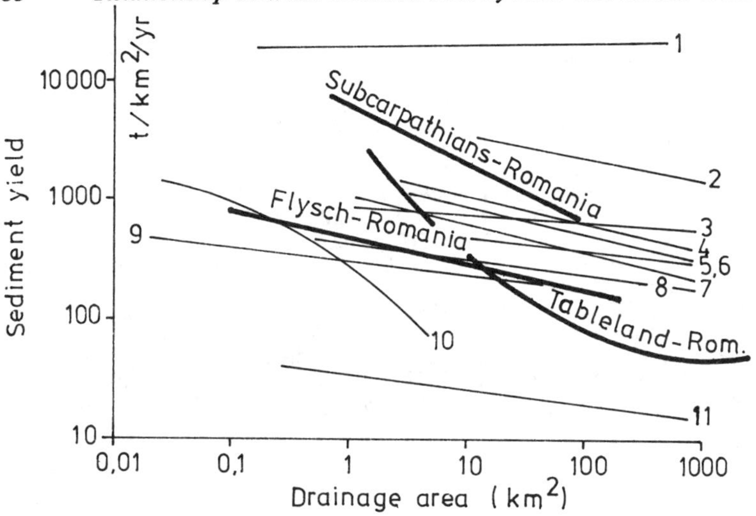

Fig. 3 Relationships between sediment yield and drainage basin area for the three study areas superimposed on those portrayed by Walling (1983) for other areas of the world.

three study areas can be attributed to contrasts in the magnitude and frequency of the discharge and in the degree of erodibility of the rock.

In situations where the rocks are easily eroded, but the discharge within the small basins is torrential in character, the sediment delivery ratio decreases markedly with an increase of the network order. This is because only during the spring flood period does the delivery system extend from the sediment sources to the outlet of the basins greater than order VII–VIII.

In the other situations more efficient delivery exists in small basins. In the Subcarpathian area, the relief, the greater amount of precipitation and the highly erodible rocks, provide a high sediment delivery ratio. In the Flysch Mountains, the volume of runoff is greater, and at the level of IVth order networks the water flow is more rapid, but the more resistant rocks are dominant and the gross erosion and the delivery ratio are both reduced. In the Moldavian hills the low contemporary delivery ratios indicate strong colluviation and glacis development at the base of the slopes.

Acknowledgements The author is grateful to Professor Des Walling from Exeter University, UK, for his helpful assistance.

REFERENCES

Arghiriade, C., Abagiu, P., Ceuca, G. & Balanica, T. (1960) Contributii la cunoasterea rolului hidrologic al padurii. *Studii si Cercetari*, 20, Bucuresti.

Baloiu, V. & Giurma, I. (1979) Schema de amenajare antierozionala a zonelor de influentă referitoare la colmatarea lacurilor de acumulare. *Bul. Inform. ASAS* **8**, 25-32.

Gaspar, R., Untaru, E., Roman, F. & Cristescu, C. (1982) Cercetari hidrologice in bazinele hidrografice torentiale mici. MFMC, Depart. Silv., ICAS.

Ichim, I. (1981) Raportul dintre eroziunea versantilor si ritmul evacuarii depozitelor dintr-un bazin hidrografic ca caza de prognoza a colmatarii lacurilor de baraj. In: *Lacuri de baraj, ambianta naturala si construita*, Academia RSR, Cluj, 129-139.

Ichim, I., Radoane, N. & Radoane, M., (1979) Elemente geomorfologice in evaluarea erozionii si acumularii in bazine torentiale amenajate din valea Bistritei. *Bul. Inform. ASAS*, **8**, 183-193.

Ichim, I. & Radoane, M. (1986) An approach to multivariate statistical analysis of the sediment yield prediction. *Proc. Workshop on Theoretical Geomorphological Models, Aachen.*

Ionita, I. & Ouatu, O. (1985) Contributii la studiul eroziunii solurilor din Colinele Tutovei. *Cercet. Agron. in Moldova*, **XIII**, 58-62.

Motoc, M., Taloescu, I. & Negut, N. (1979) Estimarea ritmului de dezvoltare a ravenelor. *Bul. Inform. ASAS*, **8**, 77-86.

Popa, A. (1977) *Cercetari privind eroziunea si masurile de combatere a acesteia pe terenurile arabile din Podisul Central Moldovenesc.* MAIA, Bucuresti.

Radoane, N. (1981) Contributii la cunoasterea unor procese torentiale din bazinul riolui Pingarati in perioada 1976-1979. *SCGGG, Seria de Geografie* **27**,(1), 53-64.

Walling, D. E. (1983) The sediment delivery problem. *J. Hydrol.* **65**, 109-237.

Erosion, Transport and Deposition Processes (Proceedings of the Jerusalem Workshop, March-April 1987). IAHS Publ. no. 189, 1990.

A simulation model for desert runoff and erosion

MIKE KIRKBY
School of Geography, University of Leeds, Leeds, LS2 9JT, UK

Abstract A distributed hillslope hydrological model is presented for a slope profile strip of variable width. It is based on stores representing soil moisture in an irregular bedrock surface, partially or entirely covered by a colluvial and/or wind-blown regolith. The model is used to explore the variations in infiltration, overland flow, soil moisture and evapotranspiration in response to differences in rainfall totals and storm durations, in gradient and in regolith depth and uniformity. Infiltration is separately estimated for bedrock and regolith components with evapotranspiration estimated from local soil moisture storage, and therefore persists longer in areas with only partial regolith cover. Biological activity is related to storage and evapotranspiration rates. Overland flow is routed downslope as a one-dimensional kinematic wave. The model is also used in aggregate form for longer periods to estimate erosion by overland flow. Some inferences are drawn about the extent to which differences between Sde Boker and the Hovav plateau are due to differences in climate, or to differences in local relief and in the supply of wind-blown sediment.

BACKGROUND: THE NORTHERN NEGEV DESERT

The work of Yair and his co-workers at sites in the Northern Negev of Israel at Sde Boqer, the Hovav plateau and near Eilat (Yair & Lavee, 1974, 1985; Schick, 1977; Yair *et al.*, 1978; Yair, 1983, 1987; Yair & Shachak, 1987) have established an impressive compilation of data for a set of desert environments, with detailed information on rainfall distributions over time and space, overland flow and sediment budgets, together with detailed surveys of plant and animal activity. Figure 1 shows the distribution of sites, loess cover, elevation and ancient agricultural fields. Inferences have been drawn *inter alia* about the influence of biological activity, and about the relationship between rainfall and sediment yield, given the pronounced increase in wind-blown loess cover in the wetter (*ca.* 150–200 mm mean annual rainfall) and less rugged (*ca.* 10 m local relief) areas of the Hovav plateau in comparison to the dryer (70–100 mm rainfall) and more rugged (*ca.* 25 m relief) Sde Boqer region.

The simulation model presented here cannot fully represent the interlocking complexity of the actual field area; but it concentrates on the slope hydrology, with a number of simplifying assumptions. The distribution of daily rainfalls is simulated as an exponential distribution, which provides a good fit for the more frequent frontal storms but may underestimate the importance

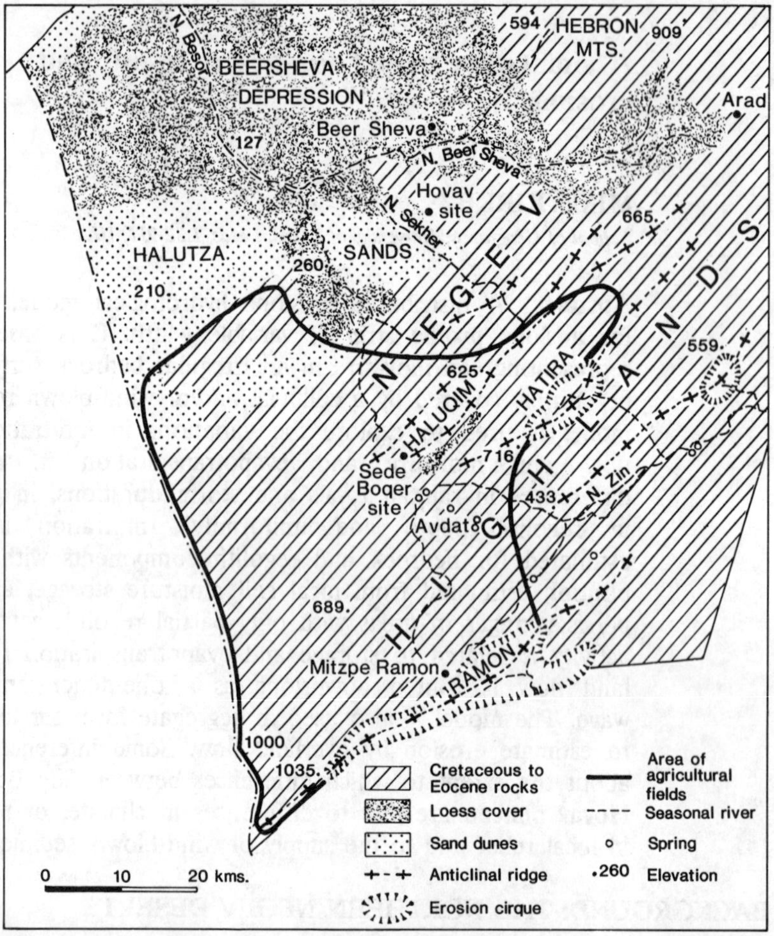

Fig. 1 Location map for the northern Negev desert, showing main rock types, areas of loess and sand cover, and area where ancient agricultural fields are found (from Yair, 1987).

of rare and intense convectional storms (Yair & Lavee, 1985). Aspects of the spatial distribution of rainfall (Sharon, 1970, 1980) have been ignored for the individual hillslope strips considered. Within each storm, the distribution of rain showers over time has been estimated from a Markov chain of independent random values. This pattern adequately simulates observed shower sequences and gives peak intensities of up to 20 mm h^{-1} over 3 min periods.

Infiltration, overland flow and evapotranspiration are estimated along a one-dimensional flow strip of variable width, so that two-dimensional flow effects and the location of rilling cannot be identified, unlike the simulation of Dunne & Aubry (1985) for Kenya. Instead the bedrock surface is envisaged as having a statistical distribution of depressions and fissures, giving an explicit relationship between mean regolith depth and proportional regolith cover.

Infiltration is first estimated for the bedrock areas. Excess rainfall from the bedrock, together with direct rainfall is considered to funnel into the regolith-filled depressions thus providing an increase in potential regolith infiltration. For the low intensity rains normally experienced in the Negev, this concentration leads to greater depths of percolation within the regolith covered proportion of largely rocky areas than occurs in areas with a complete regolith cover. Nevertheless, the total average infiltration is less in the rockier areas, and the overland flow production greater. Evapotranspiration is also estimated separately from rock and regolith proportions of the surface, and is assumed to increase towards a potential rate for high levels of moisture storage. The greater percolation depth in regolith patches leads to a longer persistence of both moisture storage and evapotranspiration. Through these mechanisms, the assumed distribution of bedrock depressions is able to mimic the hydrological responses described by Yair & Shachak (1987).

Overland flow velocities are estimated from flow depth, relying on data presented by Emmett (1978). Flow is routed down slope, taking into account differences in overland flow production resulting from surface composition, and differences in the width of the flow strip. The short duration of individual showers within each rainstorm leads to brief bursts of overland flow which rarely travel more than a few metres downslope before re-infiltrating. Where, as at Sde Boqer, the base of the slope is largely regolith covered (except immediately around the flow collector), there is a much greater overland flow discharge at rocky mid-slope sites than at regolith covered slope base sites. Where, as at sites studied on the Hovav plateau, the whole area is at least thinly mantled with regolith, overland flow is minimal. The results of simulations again mimic the differences observed at the field sites.

During a period between 80 000 and 10 000 years BP, there is evidence for loess accumulation within the Negev area (Yair, 1987). Its present distribution probably follows, to a great extent, original differences in deposition, related primarily to source areas, but also by differential washing out of wind carried dust by rainfall. The distribution may also, in part, reflect subsequent differences in erosion rates. Differential washing out and differential erosion may be partly responsible for the greater loess cover in the wetter lowland areas to the north, including the Hovav plateau site, compared to the dryer and more rugged areas to the south. The simulation model allows loess deposition, and erosion by rainsplash and wash, to proceed concurrently or in turn, so that the factors of topography, rainfall and loess deposition rates may be investigated independently. This provides some test of the inferences drawn by Yair (1987) and Yair & Shachak (1987) about the relationship between rainfall and sediment yield for the Negev area.

MODEL FORMULATION

The simulation requires formal specification of macro- and micro-topography, rainfall, infiltration, overland flow, evapotranspiration and sediment transport. These specifications are set out briefly below.

The macro-topography is based on the general features of the Sde Boqer site, and in particular on plot 2, A-C, shown in Fig. 2. The simulated slope is

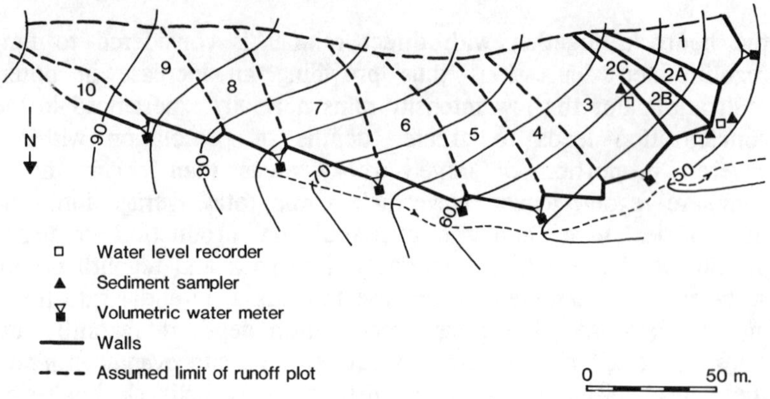

Fig. 2 Sde Boqer Field plots 1-10 (from Yair, 1987).

shown in Fig. 3, with a near uniform gradient and a linear increase in regolith cover downslope, from 0% at the divide to 70% cover at the base (Fig. 4). The general widening of the strip has been included but not the detailed form of the divide nor the basal narrowing towards the runoff collectors at the lowest point. It should also be noted that the small area of bedrock immediately surrounding the collectors has not been included in the simulation.

The micro-topography has been simulated by assuming that the proportion $p(z)$ of the area depressed more than a depth z below local highs is given by:

$$p(z) = \exp(-z/z_0) \qquad (1)$$

for a mean depth z_0, which has been taken as 250 mm in most simulations. This distribution implies the existence of some deep fissures but no high pinnacles. If this pattern of depressions is filled to an average depth z_1 with regolith material, then the proportion of soil cover is simply z_1/z_0 or 1.0, whichever is less. The depressions are then filled to within a depth of $z_0 \ln(z_0/z_1)$ of the local highs (for $z_1 <= z_0$), or the highs are buried to depth $z_1 - z_0$ (for $z_1 >= z_0$).

The distribution of daily rainfalls has also been taken as exponential, with the number of rains exceeding r given by:

$$N(r) = N \exp(-r/r_0) \qquad (2)$$

where N is the mean annual number of rain days, and r_0 is the mean rain per rain day (= Mean Ann. Rf/N). This distribution has been successfully fitted for a range of climates and time periods, provided there is not strong seasonality. For a 32 year period, the Sde Boqer site averaged 93 mm, ranging from 34 to 167 mm, falling on 15 to 42 days. For the simulation, an average of 93 mm was used over an average of 25 days per year. A random 32 year period gave a range of 54 to 170 mm, falling on 19 to 35 days, showing adequate agreement.

Within storms, the pattern of rain showers has been simulated using a

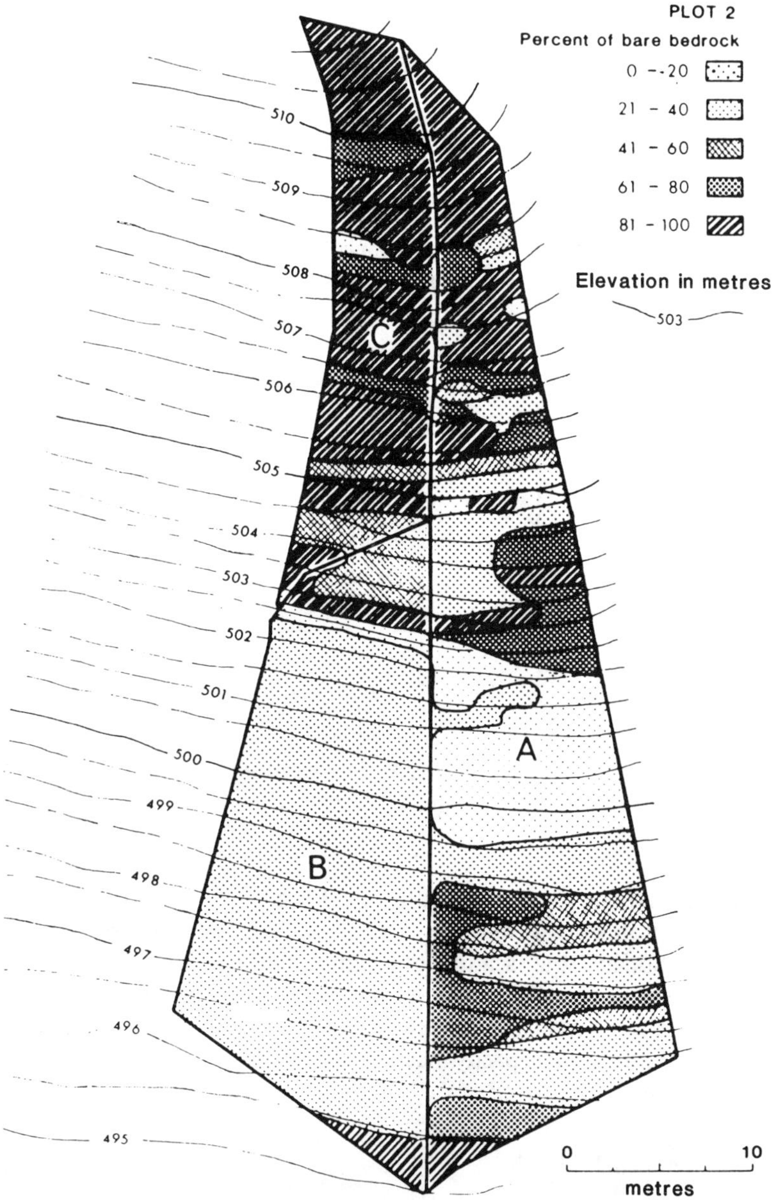

Fig. 3 Sde Boqer field plots 2 A-C, showing the proportion of bare bedrock and plot boundaries (from Yair & Lavee, 1985).

first order Markov chain with 50% persistence, so that each value is the mean of the previous value and a random variable. The random variable is obtained from a normal distribution with zero mean, using the positive values and replacing the negative values by zeros. The variance of this distribution is adjusted to give the required mean of approximately 2 mm h^{-1} overall. Figure 5 illustrates some simulated 20 mm storms showing the shower pattern produced. It may be compared with the actual storms

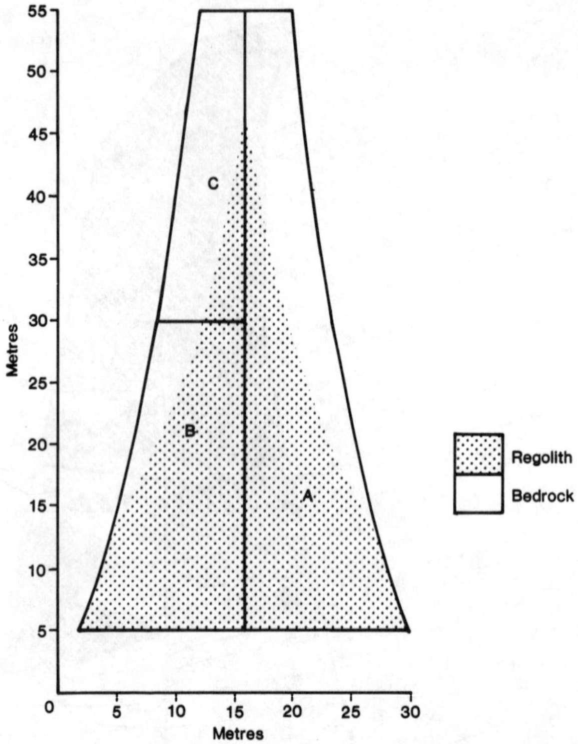

Fig. 4 Simulated flow strip map showing proportions of bedrock and regolith, modelled on plots shown in Fig. 3.

illustrated in Fig. 6. Figure 7 shows the distribution of storm amounts and durations. It may be seen that the extreme storms contained in Fig. 6 lie outside the scatter of points, showing that the simulation provides a narrower grouping than that actually observed, and that the 50% persistence should be increased.

Infiltration capacity has been calculated using the Green & Ampt (1911) equation of the form:

$$f = f_0 + B/S \qquad (3)$$

where S is a water storage level; and f_0, B are constants.

This formulation is consistent with the Philip (1957, 1958) time-based infiltration equation:

$$f = f_0 + (B/2)^{½} \, t^{-½} \qquad (4)$$

for an initially dry soil. The near-surface storage level S is increased by rainfall, and simultaneously decreased by the steady infiltration rate, f_0, which is added to a percolation store, P. Then for infiltration under rainfall at a steady intensity i (> f_0), the time to ponding is given by:

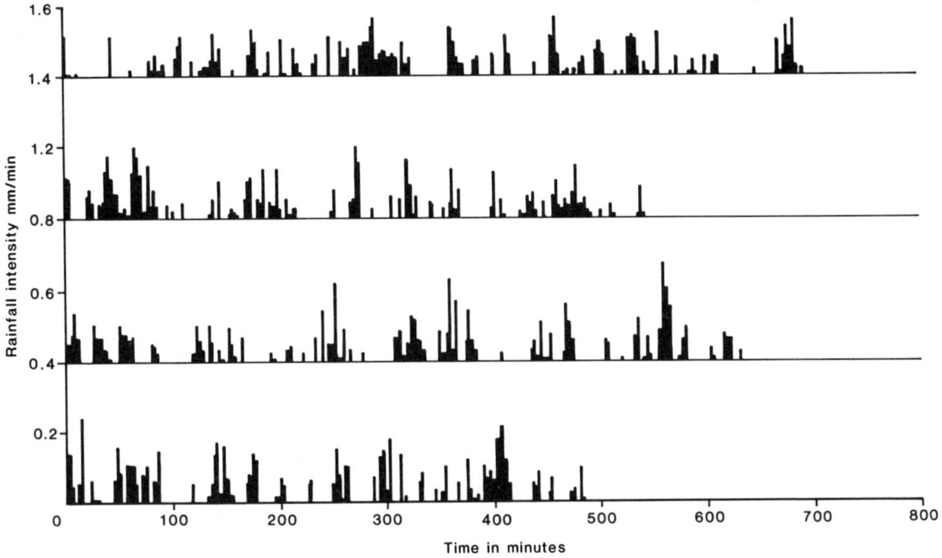

Fig. 5 Four 20 mm rain storms simulated by 3 min units.

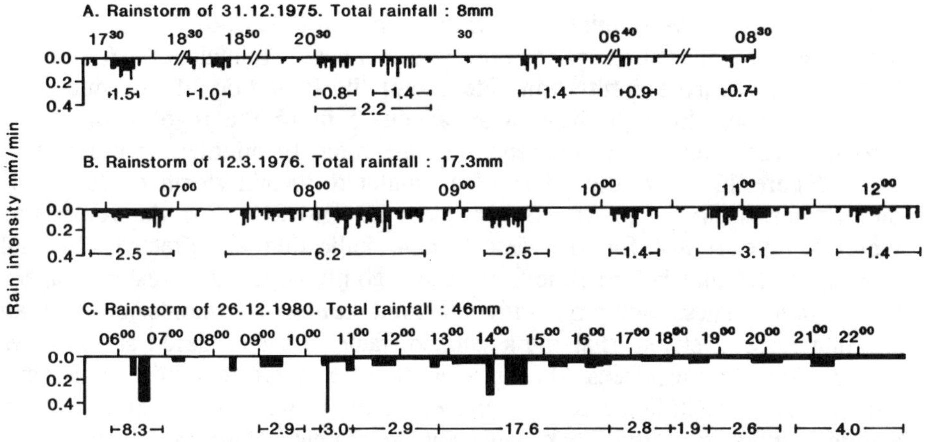

Fig. 6 Three natural frontal rain storms for the Sde Boqer area (from Yair, 1987).

$$t_0 = B/(i - f_0)^2 \tag{5}$$

The values used in this simulation are:

	f_0(mm h^{-1})	B(mm^2 h^{-1})
Bedrock	0	2.4
Regolith	10	24

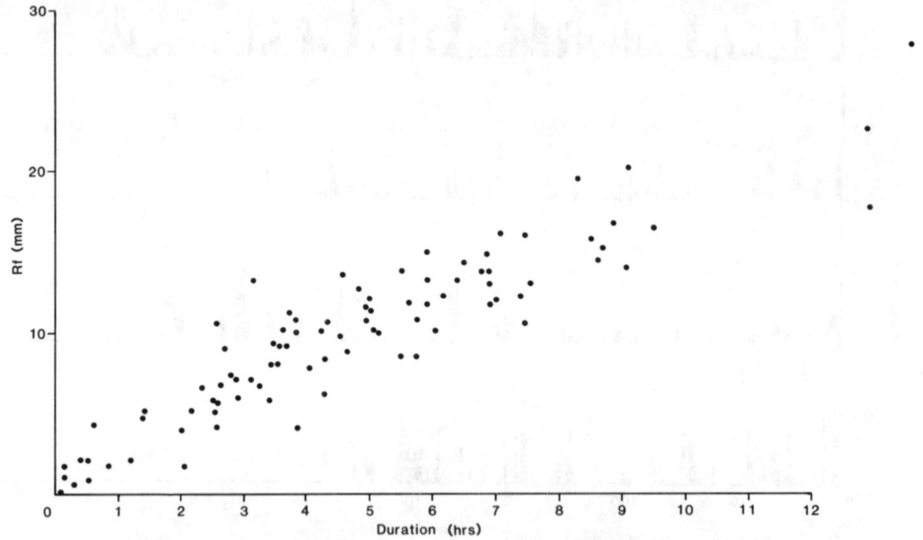

Fig. 7 The scatter of simulated rain storm amounts and durations.

Figure 8 shows simulated infiltration rates for bedrock and regolith. In comparison to measured rates under a simulated rainfall of 26 mm h^{-1}, (Fig. 9), the bedrock infiltration rate shows the limitation of an equation of the above type, although there is an adequate fit to the regolith data. The forecast initial rate is too low, and the rate after 10 minutes or more is too high. Figure 10 shows the effect of a simulated 30 min storm of 26 mm h^{-1} on flow strips similar to Sde Boqer, plots 2B and 2C. For the rocky slope there is little runoff for the first 3 min, indicating an effective threshold capacity of 1.3 mm before runoff; for the regolith slope the threshold capacity is 6.5 mm. These compare with observed runoff thresholds of 1–3 and 3-5 mm respectively, so that the simulated values are considered satisfactory.

In low intensity rains, the amount of water infiltrated into areas which are partially regolith-covered is higher than a weighted average of the amounts infiltrated into rock and soil separately. This is because water running off the rock is allowed to funnel into the regolith areas and infiltrate. Nevertheless the total infiltration is less than for areas completely covered with regolith, and the overland flow production is greater.

Overland flow velocities have been estimated using Emmett's (1970) data, although these show little pattern other than a general increase of mean velocity with flow depth. The theoretical exponent of depth to give flow velocity should be 2/3 for turbulent flow and 2 for laminar flow. The data suggest that a value of 1.0 provides a tolerable fit, and that there is little evident dependence on slope gradient. The relationship:

$$v = 10\, d \qquad (6)$$

has therefore been adopted, where v is the mean velocity in m s^{-1} and d is the mean flow depth in m. The kinematic wave velocity should be exactly

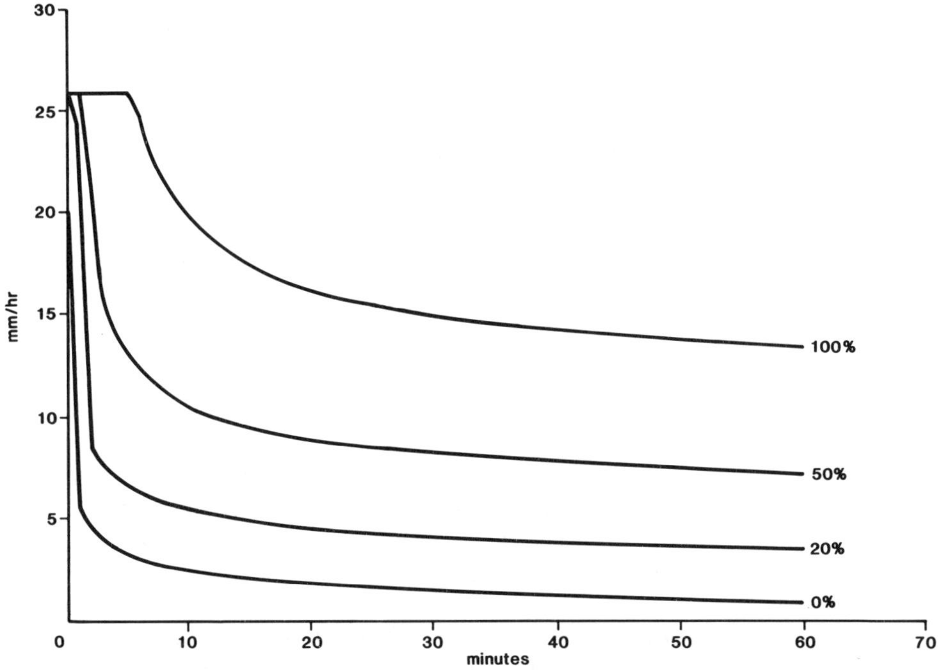

Fig. 8 Simulated infiltration rates during rainfall at 26 mm h^{-1} into slopes of differing regolith cover.

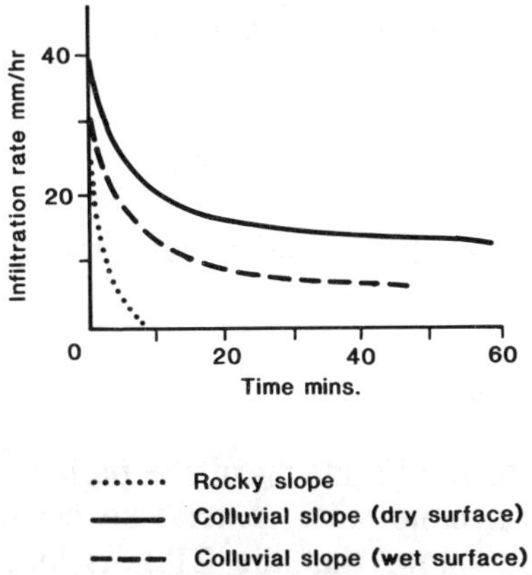

Fig. 9 Measured infiltration rates during artificial rainstorm at 26 mm h^{-1} on bedrock and colluvium covered surfaces (from Yair & Shachak, 1987).

twice this value. For a flow of 1 l s^{-1} m^{-1}, this relationship gives a flow depth of 10 mm at a mean velocity of 0.1 m s^{-1}.

Overland flow has been routed downslope along the variable width flow strip as a kinematic cascade, using 3 min iterations over 2.5 m distance increments until flow and rainfall come to an end. This part of the model is very similar in principle to that described by Yair & Lavee (1985). For the

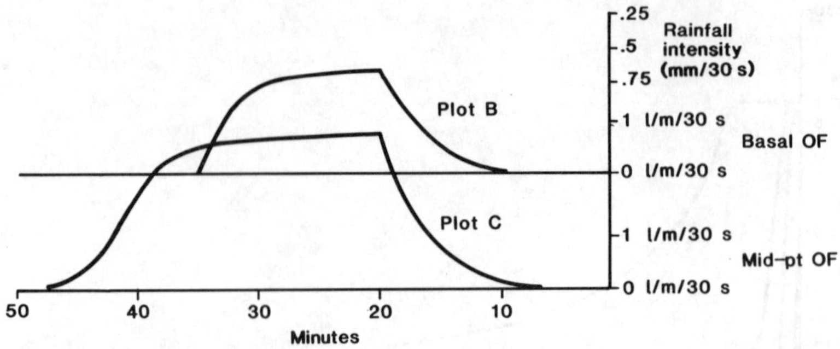

Fig. 10 Simulated runoff from plots B and C for a 30 min rainstorm at 26 mm h^{-1}.

remainder of each rain day, and for days without rainfall, evaporation is integrated over the whole of the relevant period. For convenience, and without significant loss of accuracy for an arid area, rain storms are invariably simulated at the start of each water day on which they occur.

Evapotranspiration (ET) is estimated for bedrock and regolith proportions of the surface at each point. It is calculated for each 3 min increment during storm events and integrated over the remainder of rain days, and for dry days as a whole. Actual ET is estimated from a potential rate e_p (assumed to be 5 mm per day). Losses are taken at the potential rate from any overland flow layer and from the near-surface store until both are exhausted. Subsequently, the actual rate is taken as a function of the percolation storage P, at rate:

$$e_p\,[1 - \exp(-P/P_0)] \tag{7}$$

where P_0 is a scale depth for the rate of attenuation. Integrating this over a period Δt, the total loss is:

$$P - P_0 \ln\{1 + [\exp(P/P_0) - 1]/\exp(e_p\Delta t/P_0)\} \tag{8}$$

where P is the percolation storage at the beginning of the period.

Because, in low intensity rains, more water enters the percolation store in areas of partial regolith cover, it follows that the ET lasts longest at these sites, and is greater per unit area of regolith. An even more favourable environment is provided at sites where regolith cover is increasing downslope, especially close to a sharp boundary, because overland flow generated upslope re-infiltrates preferentially, so that the total ET and not only the ET per unit area of regolith, is increased. In favourable sites, simulations suggest that the total ET may be 25–35% greater than the rainfall.

Although biological components are not explicitly included in this model, an estimate of density for vegetation cover, and for soil fauna depending on it may be obtained as follows. Suppose the mean actual ET is AE in an area with a proportion p of regolith cover; and that the required ET per unit area to support a vegetation type is RE (perhaps 25% of the potential ET).

Accordingly, in the area of adequate unit ET, that is where $AE/p > RE$, the relative density is proportional to p. In the area of insufficient ET however, the density is proportional to $p[AE/(pRE)]^m$, for an exponent $m > 1$ (for the sake of illustration let $m = 4$). A power law gives continuity at the cross-over point, at which the regolith proportion p is equal to the ratio AE/RE, and expresses the probability of a pocket of locally concentrated soil moisture.

The final component of the model is sediment transport by wash and splash processes. Rainsplash is estimated as proportional to the square of the rainfall intensity, and is considered to occur only on regolith covered parts of the surface. Wash is estimated as proportional to the square of the overland flow discharge, and a relationship of this kind is supported by some sediment data for the Sde Boqer site (Fig. 11). In this case the concentration of regolith areas in depressions has two opposing influences. First, it may lead to flow concentration within the depressions, depending partly on their orientation; and second the depressions may provide local reverse slopes which hinder sediment evacuation, particularly for low proportions of regolith cover, at which the regolith material is concentrated in fissures. As a compromise the rate of wash transport has been taken as proportional to the square root of proportional regolith cover, so that large amounts of cover give little attenuation, and *vice versa.*

Both splash and wash have been estimated as directly proportional to gradient, and no lower threshold for wash has been set, as seems appropriate for wind deposited material (although a substantial threshold would be

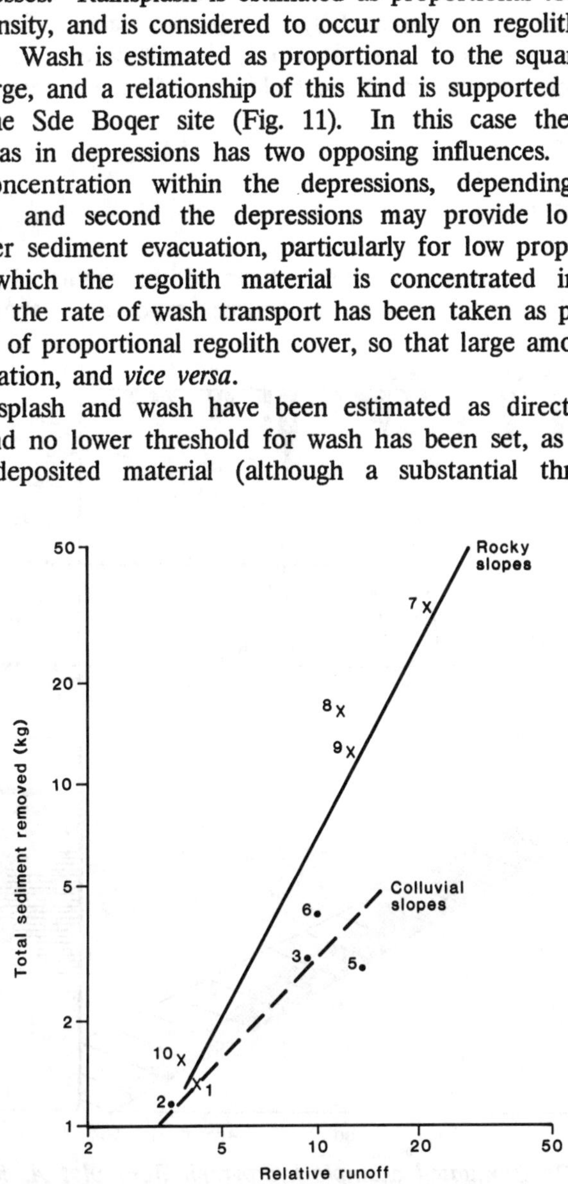

Fig. 11 Total sediment transport and total plot runoff (as % of combined plot runoff) for Sde Boqer plots, for the year 1972–1973 (data from Yair & Shachak, 1987).

appropriate for stony colluvial material). A balance between wash and splash rates has been arbitrarily chosen to give dominance to wash in areas of significant overland flow. The balance between these processes and rates of sediment supply, either as wind-blown loess or by weathering of bedrock as a source of colluvium, has been varied in order to examine the effect of differing ratios of supply to removal in the location of the regolith cover. In order to span from the hydrological simulations of a few years up to relevant periods for erosion and deposition of a few thousands of years, each year's sediment yields have been multiplied by a replication factor of 50 to 200 to provide detectable changes in the distribution of soil cover. Even for these time spans however, no erosion of the bedrock surface has been included in the simulation.

SIMULATION RESULTS

Figure 12 illustrates a number of the features of the simulated hydrographs and sediment yield for a 50 m plot with bedrock exposed for the top quarter

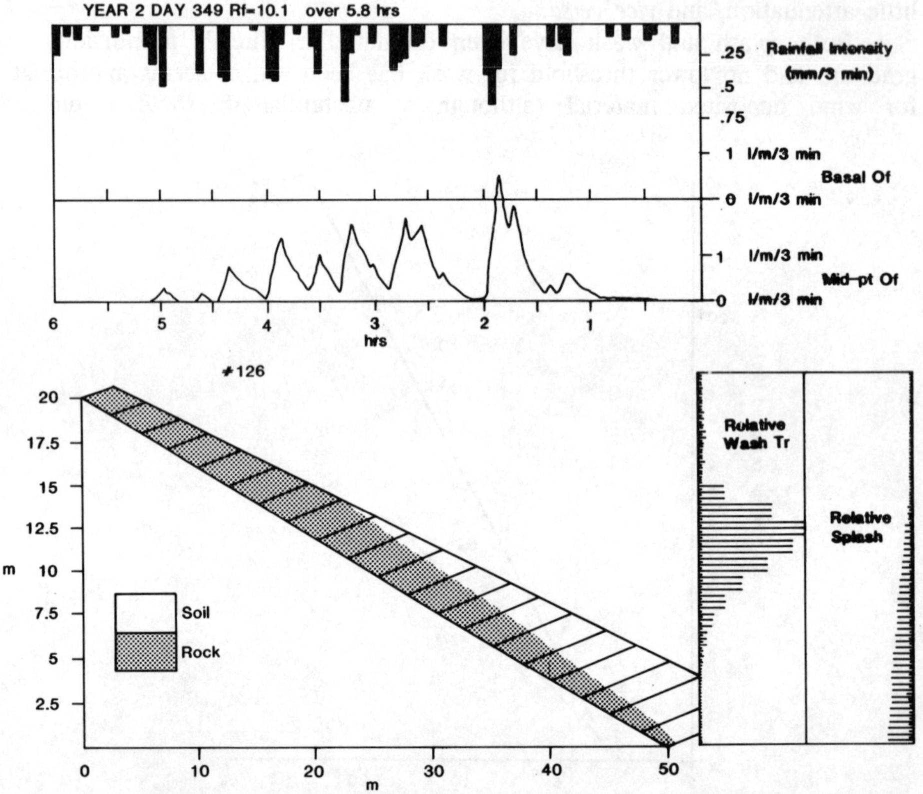

Fig. 12 Simulated example hydrograph from plot A, for the slope profile form shown. The relative cumulative magnitudes of sediment transport by rainsplash and wash down the slope length are also shown.

of the slope, and regolith cover then linearly increased to 90% at the slope base. It may be seen that there is a substantial response to rainfall at mid-slope sites, but little from the regolith areas at the foot of the plot. In comparison with Sde Boqer, the mid-slope flows are of the correct order of magnitude; but the slope base flow is much less than from the field plot, probably because of the rocky area immediately above the slope-base collector. The rain showers are brief enough, even on the relatively short 50 m slopes, and show an appreciable increase in flow downslope only within rocky areas. Figure 13 shows annual overland flows for the configuration shown in Fig. 12, with a pronounced peak at the foot of the rocky area. The accumulation downslope may be compared with flows along a completely rocky slope, where about 70 m is required to reach equilibrium discharge, and on a slope with 15% regolith cover where greater infiltration produces equilibrium within about 20 m as well as greatly reduced total flows.

Figure 12 also shows the pattern of sediment transport rates forecast from the flow and rainfall data for this profile of almost uniform gradient. It may be seen that splash rates increase steadily downslope, largely reflecting the increased proportion of regolith cover downslope. Wash rates, on the other hand, show a marked peak (at 22 m from the crest), somewhat downslope from the overland flow peak (at 17 m) because of the increasing availability of sediment in a downslope direction.

Over a period, the effect of this sediment transport peak is to "sweep" sediment downhill, especially in the area near the top of the regolith. Figure

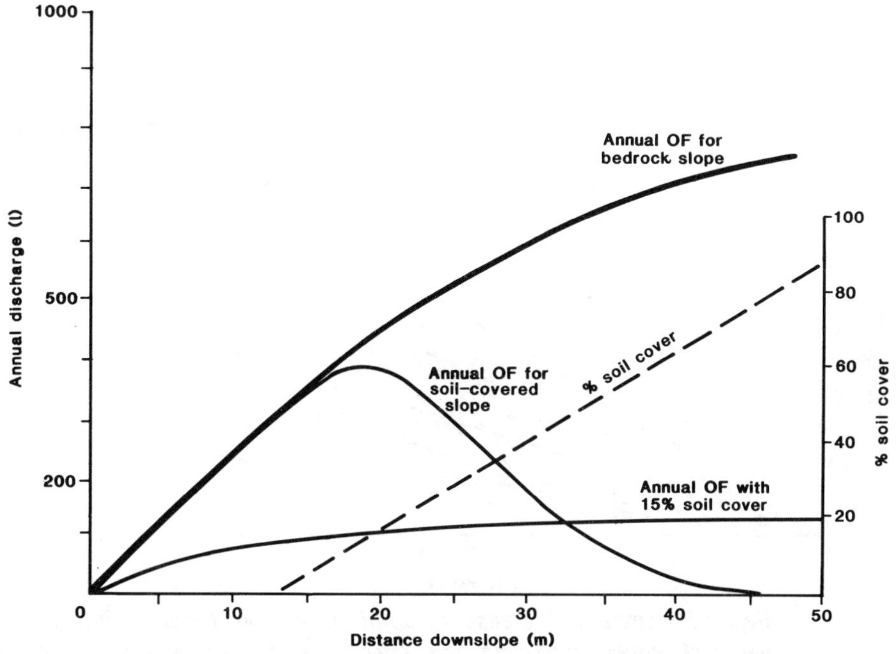

Fig. 13 Simulated annual overland flow discharge for slopes of constant regolith cover, and for slopes with regolith cover increasing downslope as indicated.

14 shows a simulation in which an initially bedrock-surfaced slope is subjected to loess deposition at a rate increasing from zero at the hill crest to 60 mm per 1000 years at the slope base. It may be seen that a thin, patchy equilibrium cover is established progressively from the crest downwards. Near the slope base, given the basal conditions used, net deposition is close to the input rate. In the middle third of the slope, however, there is a transition, which becomes more abrupt with time, from a mainly bedrock slope to a mainly regolith covered slope.

In this way, sediment can be effectively swept downslope in areas where there is a balance between loess input and sediment transport capacity. In areas of relatively higher loess input, the sweep effect is slight and the sweep zone migrates upslope, whereas in areas of relatively lower loess input, the sweep zone migrates downslope until, in extreme cases, the entire deposit is removed. In the transect from north to south across the northern Negev, the field evidence thus suggests that the influence of high deposition rates and low gradients in the north outweighs the effect of greater rainfall and allows loess to cloak the landscape. In the south, the low deposition rates and steep gradients similarly appear to be more important in preventing loess accumulation than the influence of low rainfall in reducing erosion. At Sde Boqer, the combination of rocky hillslopes and loess-filled valleys appears to be much closer to the critical balance.

The effect of erosion sweeping of loess can also be seen in annual ET and storage estimates. Figure 15 shows the same time sequence as in Fig. 14 for mean ET rates. There is some slight scatter produced by using data from

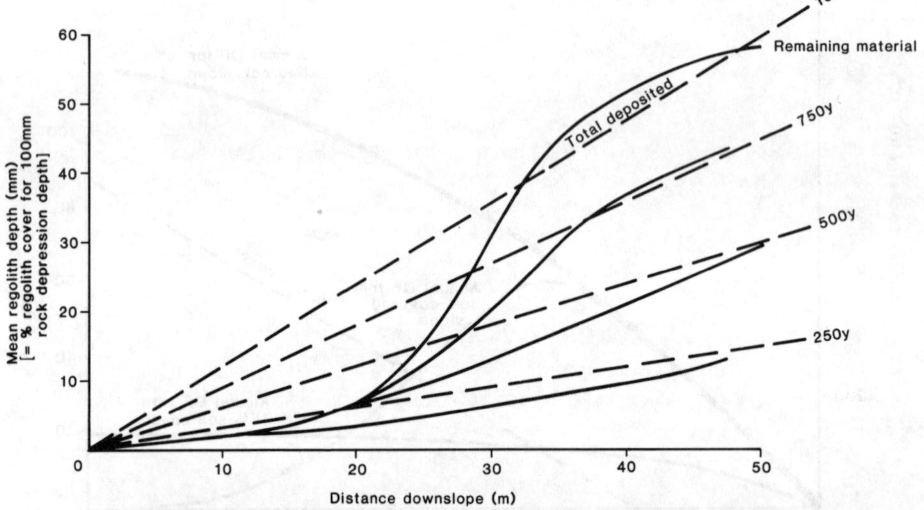

Fig. 14 Simulated course of total deposition (straight lines) and net soil cover, including the effects of deposition and sediment transport. Loess deposition increases linearly downslope to 60 mm per 1000 years at the slope base. The effect of sediment sweeping may be seen.

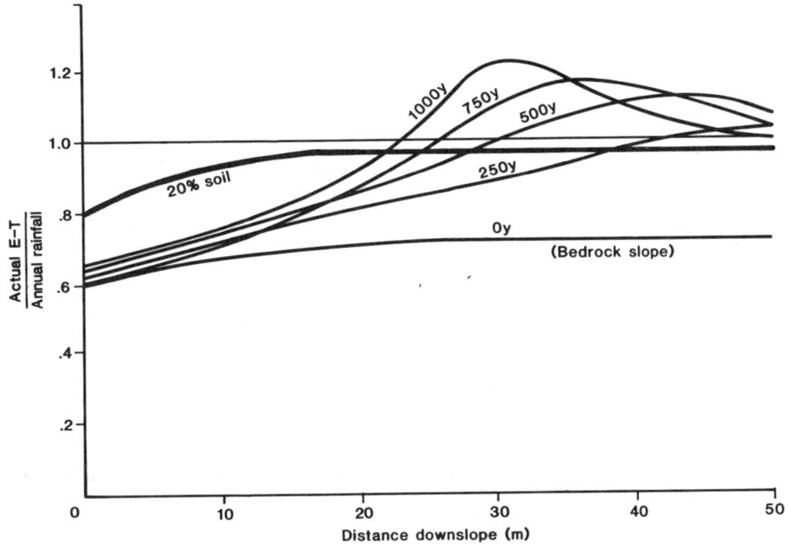

Fig. 15 Enhancement of simulated mean annual evapotranspiration by erosion "sweeping". The time sequence refers to the same run as in Fig. 14.

particular years with higher or lower total rainfall, but a clear pattern is evident. For comparison the influence of uniform loess deposition is also indicated. The sweep effect is responsible for bringing mean rates above the rainfall in the zone of declining overland flow. Overland flow re-infiltrates into the loess in the sweep zone, producing a progressively sharper peak in soil water storage and ET. In this example the mean ET is about 20% greater than the rainfall, and in some runs this enhancement rises to over 30%.

Some estimate of relative plant density may be inferred from these evapotranspiration data as described above. In Fig. 16, the same sequence has again been used for illustration. Local ET in rainfall patches has been estimated as the ratio of mean ET to proportion of rainfall cover, and a critical value of 450 mm (approximately 25% of the assumed annual potential ET of 1825 mm) has been arbitrarily chosen as the threshold level for the growth of perennial plants and the establishment of associated herbivores. In the area of inadequate local ET, a fourth power law has been used to sketch the decline in density. Slopes of constant loess cover have also been included for comparison. The densities referred might best be identified, *a priori*, with root spread densities: crown covers are likely to be half as great, or even less.

In the simulation, the most favourable uniform slope for biological activity appears to be one with about 20% regolith cover, with sharp reductions in density on either side of this optimum. During the simulated evolution of the regolith cover, as sediment is swept downslope, there is a narrow zone of maximum biological activity, which may be locally richer than for the best uniform slope. On either side of this maximum, biological activity declines sharply; upslope, because of a shortage of regolith sites, and

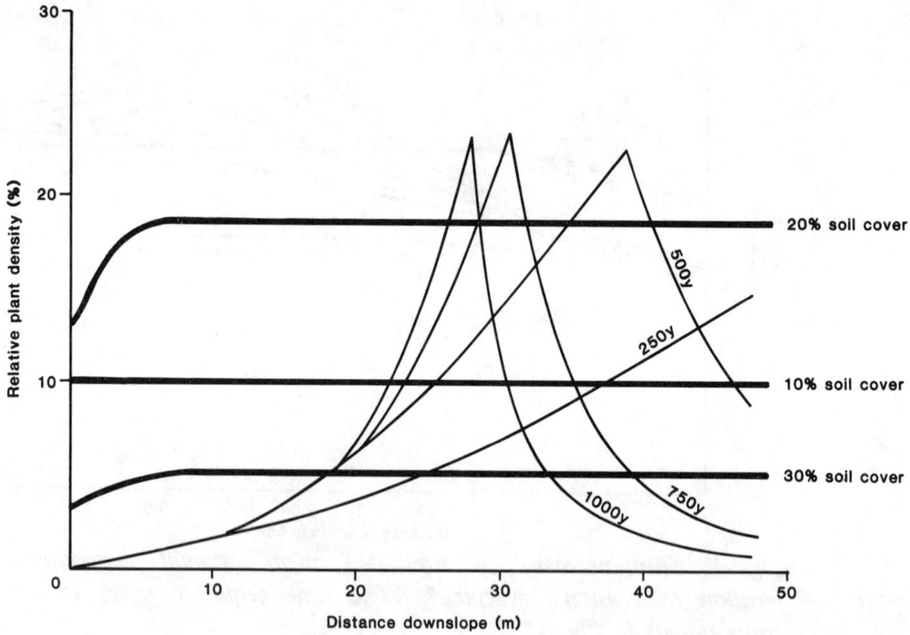

Fig. 16 Estimated relative plant densities (perhaps related to root cover), estimated from simulated hydrological data, for uniform rainfall covers and for the erosion and deposition sequence shown above.

downslope because of a lack of water within rooting depth. When the distribution of annual rainfalls is taken into account, the sharpness of this peak is somewhat softened, so that the apparent local advantage may be less narrowly confined. It may also be seen that as soil cover increases over time, the breadth of the favoured area for plants tends to become narrower as it migrates upslope.

CONCLUSIONS

Some tentative conclusions may be drawn from the simulation about the relationship between sediment yield and rainfall in this arid environment, and about the nature of the relationship between biological activity and sediment yield.

In estimating sediment yields, no account has been taken of the possible effects which are commonly thought to control Langbein & Schumm's (1958) well-known relationship. In particular, no account is taken of the influence of vegetation crown cover in shielding the surface from the effects of rainsplash and crusting, nor of improved infiltration capacity in response to increases in soil organic matter. In addition, no consideration has been given to increased flow resistance from vegetative roughnesses. For the densities considered in the simulation, and observed in the field, it is argued that these effects can be largely ignored in the range of rainfalls and vegetation covers involved, at least up to about 200 mm annual rainfall. The maximum root density is estimated, for optimum loess cover, at about 40%, with corresponding crown cover of 20% or less. Given the short showers which make up most of the rainfall, the effect of increased rainfall should be an almost linear increase in both wash and splash erosion, with the main difference in the frequency of flows rather than in their peak discharge. Thus the overall response to increasing rainfall, for a given degree of loess cover, is seen as a clear, and almost linear, increase in sediment yield. The most important possible counter argument is that the rate of loess deposition is causally linked to rainfall, through scrubbing out of wind blown dust by the rain.

Another conclusion from the simulation is that plant and animal densities, as well as the spatial pattern of sediment transport, can both be explained as depending on the slope hydrology. Figure 17 shows a fairly clear empirical relationship between actual sediment yield and biologically available material, but it is plain, from the contrast between plots with a colluvial footslope and those in bedrock, that other and presumably hydrological factors are at least as important as biological disturbance. It is therefore argued that there need not be a cause and effect relationship between biological activity and erosion, but that both may primarily be independent responses to the hydrological regime.

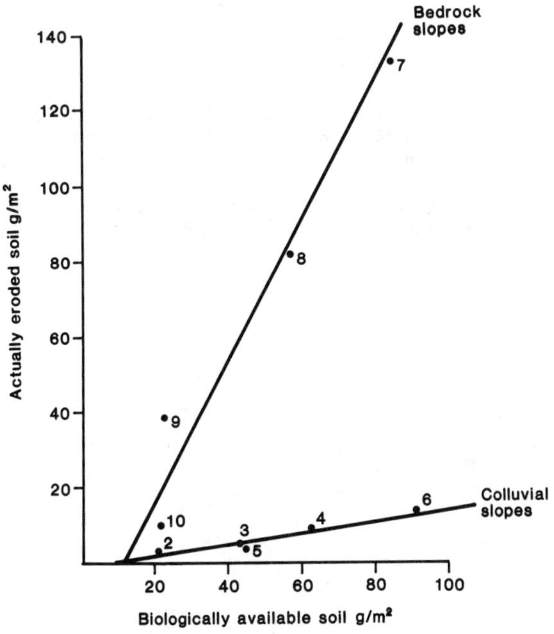

Fig. 17 Empirical relationship between soil actually eroded and biologically "available" material, for the 8 mm storm of 24 November 1972 on Sde Boqer plots (data from Yair & Shachak, 1987). Note the strong distinction between bedrock plots and those with a colluvial base.

Acknowledgement I would like to thank Aaron Yair for this advice and support, and for providing the data which are referred to in this paper.

REFERENCES

Dunne, T. & Aubry, B. F. (1985) Evaluation of Horton's theory of sheetwash and rill erosion on the basis of field experiments. In: *Hillslope Processes* (ed. A. D. Abrahams), 31-53. Allen & Unwin, Boston, Massachussetts, USA.

Emmett, W. W. (1970) The hydraulics of overland flow on hillslopes. *USGS Prof. Pap.* **662A**.

Emmett, W. W. (1978) Overland Flow. In: *Hillslope Hydrology* (ed. M. J. Kirkby) 145-176. John Wiley, Chichester, UK.

Green, W. H. & Ampt, G. A. (1911) Studies in Soil Physics. 1: The flow of air and water through soils. *J. Agric. Sci.* **4** (1), 1-24.

Langbein, W. B. & Schumm, S. A. (1958) Yield of sediment in relation to mean annual precipitation. *Trans. AGU* **39**, 1076-1084.

Philip, J. R. (1957, 1958) The theory of infiltration. *Soil Sci.* **83**, 345-357 & 435-448; **84**, 163-177, 257-264 & 329-339; **85**, 278-286 & 333-337.

Sharon, D. (1970) Areal patterns of rainfall in a small watershed. In: *Symposium on the Results of Research on Representative and Experimental Basins,* (Wellington 1970), Vol. I, IAHS Publ. no. 96, 3-11.

Sharon, D. (1980) The distribution of hydrologically effective rainfall incident on sloping ground. *J. Hydrol.* **46**, 165-188.

Schick, A. P. (1977) A tentative sediment budget for an extremely arid catchment in the southern Negev. In: *Arid Geomorphology* (ed. D. O. Doehring), 139-163. Publications in Geomorphology.

Yair, A. (1983) Hillslope hydrology, water harvesting and areal distribution of ancient agricultural fields in the northern Negev desert. *J. Arid Environ.* **6**, 283-301.

Yair, A. (1987) Environmental effects of loess penetration into the northern Negev desert. *J. Arid Environ.* **13**, 9-24.

Yair, A. & Lavee, H. (1974) Areal contribution to runoff on scree slopes in an extreme arid environment: a simulated rainstorm experiment. *Z. Geomorphol. Suppl. Bd.* **21**, 106-121.

Yair, A. & Lavee, H. (1985) Runoff generation in arid and semi-=arid zones. In: *Hydrological Forecasting* (ed. M. G. Anderson & T. P. Burt), 183-220. John Wiley, Chichester, UK.

Yair, A. & Shachak, M. (1987) Studies in watershed ecology of an arid area. In: *Progress in Desert Research* (ed. L. Berkofsky & M. G. Wurtele), 145-193. Rowman & Littlefield, Totowa, New Jersey, USA.

Yair, A., Sharon, D. & Lavee, H. (1978) An instrumented watershed for the study of partial area contribution of runoff in the arid zone. *Z. Geomorphol. Suppl. Bd.* **29**, 71-82.

Erosion, Transport and Deposition Processes (Proceedings of the Jerusalem Workshop, March-April 1987). IAHS Publ. no. 189, 1990.

Spatial variability of overland flow in a small arid basin

HANOCH LAVEE

Department of Geography, Bar-Ilan University, Ramat Gan, Israel

AARON YAIR

Department of Physical Geography, Hebrew University of Jerusalem, Jerusalem, Israel 91904

Abstract Overland flow generation and continuity along hillslopes in a small arid drainage basin in the northern Negev, Israel, has been assessed on the basis of field measurements. Variables measured included rainfall, overland flow, evaporation, infiltration rates and soil moisture. Data obtained show great spatial and temporal variability among all variables. The most frequent phenomenon encountered was overland flow discontinuity along slopes having lengths ranging from 55 to 76 m. This discontinuity is attributed to the short duration of rain showers and can be enhanced by surface properties (where the infiltration rate increases downslope). A deterministic simulation model for the spatial and temporal variations in overland flow along the hillslopes was developed taking into consideration the spatial and temporal distribution of the variables mentioned above. The model is based upon functions derived from the field measurements and permits the delineation of the hillslope area contributing to channel flow under various rain conditions. Results obtained may explain the use of certain water-harvesting techniques adopted by farmers living in the area some 2000 years ago.

INTRODUCTION

Arid and semi-arid areas are known for the frequent occurrence of Hortonian overland flow (Horton, 1945). Recent studies (Yair *et al.*, 1978, 1980; Lavee, 1982) have shown that overland flow generation is highly non-uniform even within a very small basin extending over 0.02 km^2. These studies drew attention to the phenomenon of flow discontinuity occurring along slopes of lengths 55 to 76 m, and attributed it to pronounced differences in infiltration rates between the upper rocky and the lower soil-covered slope sections.

The possibility however, that overland flow discontinuity can frequently occur over uniform slopes, or even over slopes having a soil-covered upper part and a rocky lower part, has not previously been investigated. Such occurrences may be related to rainfall characteristics rather than to surface properties. If so, overland flow discontinuity along arid hillslopes is not merely a local phenomenon but is in fact general. This has important theoretical and practical applications.

This paper has two main objectives:

(a) to analyse the generality of overland flow discontinuity over arid slopes having different surface properties, and

(b) to apply the phenomenon of flow discontinuity over hillslopes towards an improved understanding of water-harvesting techniques used by the ancient farmers that lived in the region 2000 years ago.

STUDY AREA AND EXPERIMENTAL DESIGN

The experimental site, situated in the northern Negev desert near Sde Boqer, has an area of 0.011 km^2 consisting of the north-facing side of a first order drainage basin (Fig. 1). Slope length varies from 55 to 76 m and slope gradient from 12 to 29.5%.

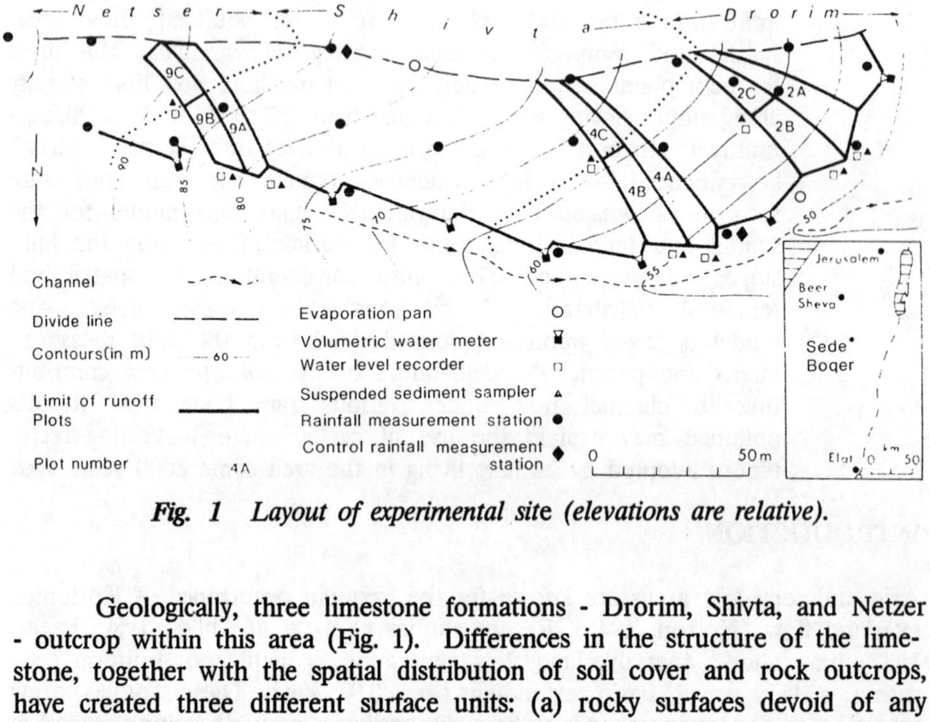

Fig. 1 Layout of experimental site (elevations are relative).

Geologically, three limestone formations - Drorim, Shivta, and Netzer - outcrop within this area (Fig. 1). Differences in the structure of the limestone, together with the spatial distribution of soil cover and rock outcrops, have created three different surface units: (a) rocky surfaces devoid of any soil cover; (b) stony soil-covered areas with limited rock outcrops; and (c) intermediate areas with varying rock/soil-cover ratios. Both the Shivta formation and the upper part of the Drorim formation belong to the first surface unit, with narrow and shallow soil-covered strips at the base of rock terraces (Fig. 2). The lower part of the Drorim formation is typical of the second surface unit whereas the Netzer formation forms the third surface unit.

Average annual rainfall in the area is 93 mm. Local rainfall measurements were taken at 21 points (Fig. 1). At each point, twin raingauges were placed 30 cm above the ground; one with a horizontal orifice and the other

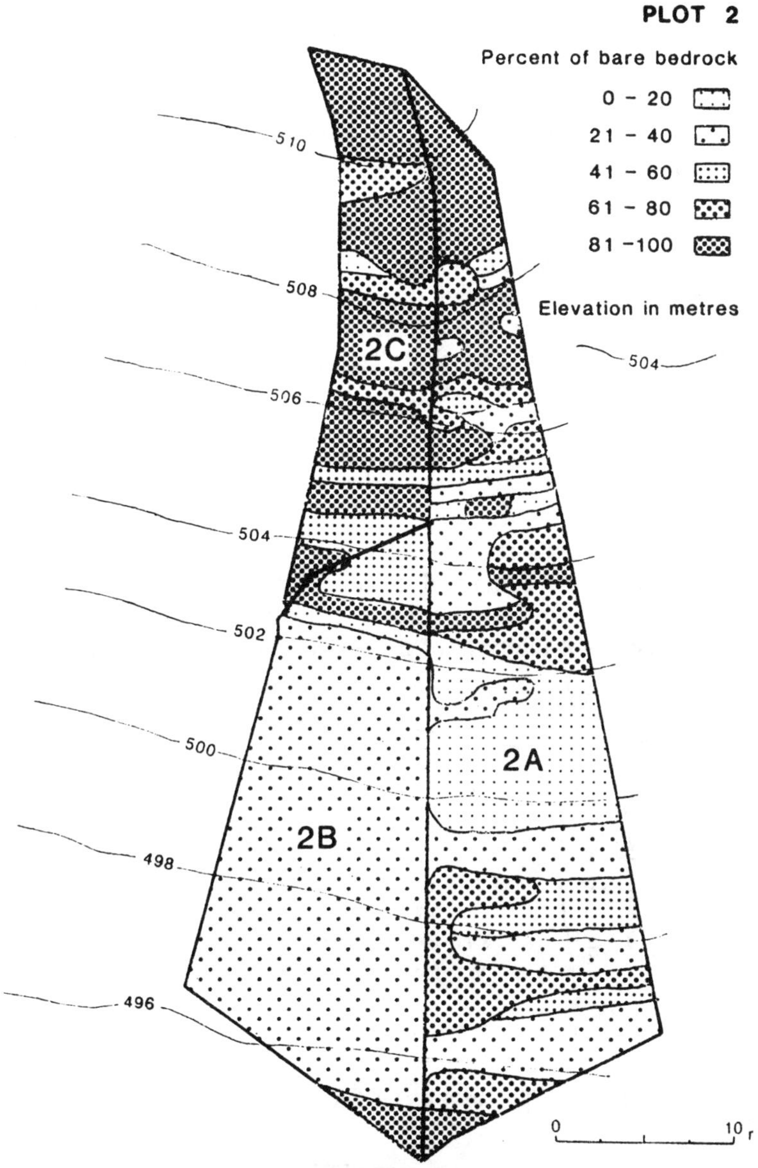

Fig. 2 Percent of bare bedrock at plot 2.

tilted such that its orifice was parallel with the ground. The former measures rainfall in terms of rainfall flux through a horizontal plane and is therefore appropriate for meteorological purposes. The latter, by taking slope inclination and aspect into account, measures the amount of rainfall actually reaching the ground. The second method is suited to hydrometeorological research where rain is an input into the system on the earth's surface (Sharon, 1980). In order to check the relationship between the rain amount at 30 cm and that reaching the ground, raingauges with their orifices at ground level were installed at two control rainfall measurement stations. Rain recorders and evaporation pans installed at these two stations provided information on rain intensity, duration, and on potential evaporation (Fig. 1).

Infiltration rates, representative of the three surface units, have been measured using a rainfall simulator (Morin *et al.*, 1970). Direct measurements of soil moisture within the upper 3 cm of the soil were taken at several points along selected slopes immediately after each rainstorm, and on the first, third, sixth, tenth, etc., days that followed.

Analysis of overland flow generation and continuity was based on three runoff plots, each subdivided into three sub-plots: one long, extending from the divide to the slope base, and two adjoining short ones draining the upper and lower sections of the slope, respectively. Each sub-plot was equipped with a water level recorder (Fig. 1).

RESULTS AND DISCUSSION

Rainfall measurements revealed that:

(a) Rainstorms are composed of series of short showers with most showers lasting less than 30 minutes (Fig. 3) and yielding less than 3 mm (Fig. 4). There are intervals of at least 6 min between showers.

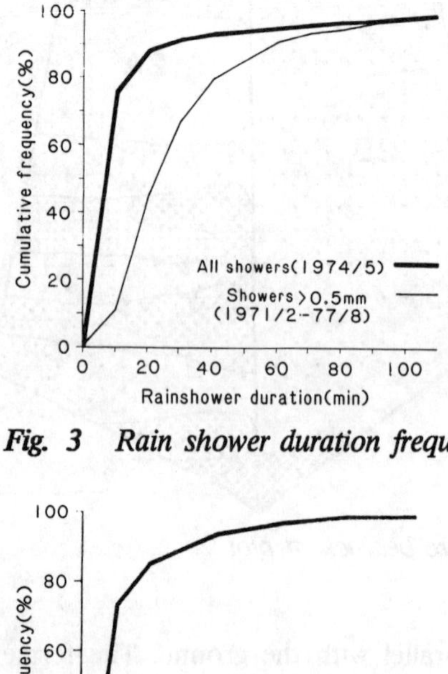

Fig. 3 Rain shower duration frequency.

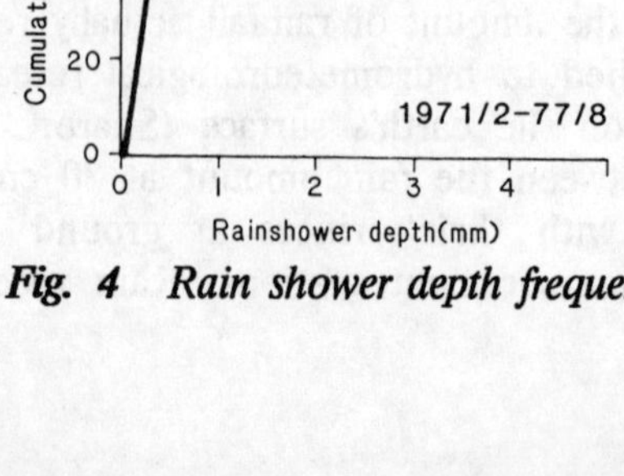

Fig. 4 Rain shower depth frequency.

(b) The spatial distribution of the hydrological rainfall, as measured by the tilted raingauges, differs greatly from the meteorological rainfall distribution, as measured by the horizontal orifice raingauges (Fig. 5). Differences in rainfall at any point varied from 10 to 40%. In measuring hydrological rainfall, differences of 20 to 80% were found to exist over distances of less than 100 m. As rainstorm duration was the same for the study area as a whole, the spatial differences in the rainfall depth represent differences in rainfall intensity.

(c) There is a constant ratio between rainfall depth at 30 cm above the ground and at ground level. This ratio is close to unity, but at the slope base it is slightly lower (1.07) than for the ridge (1.13) where the wind

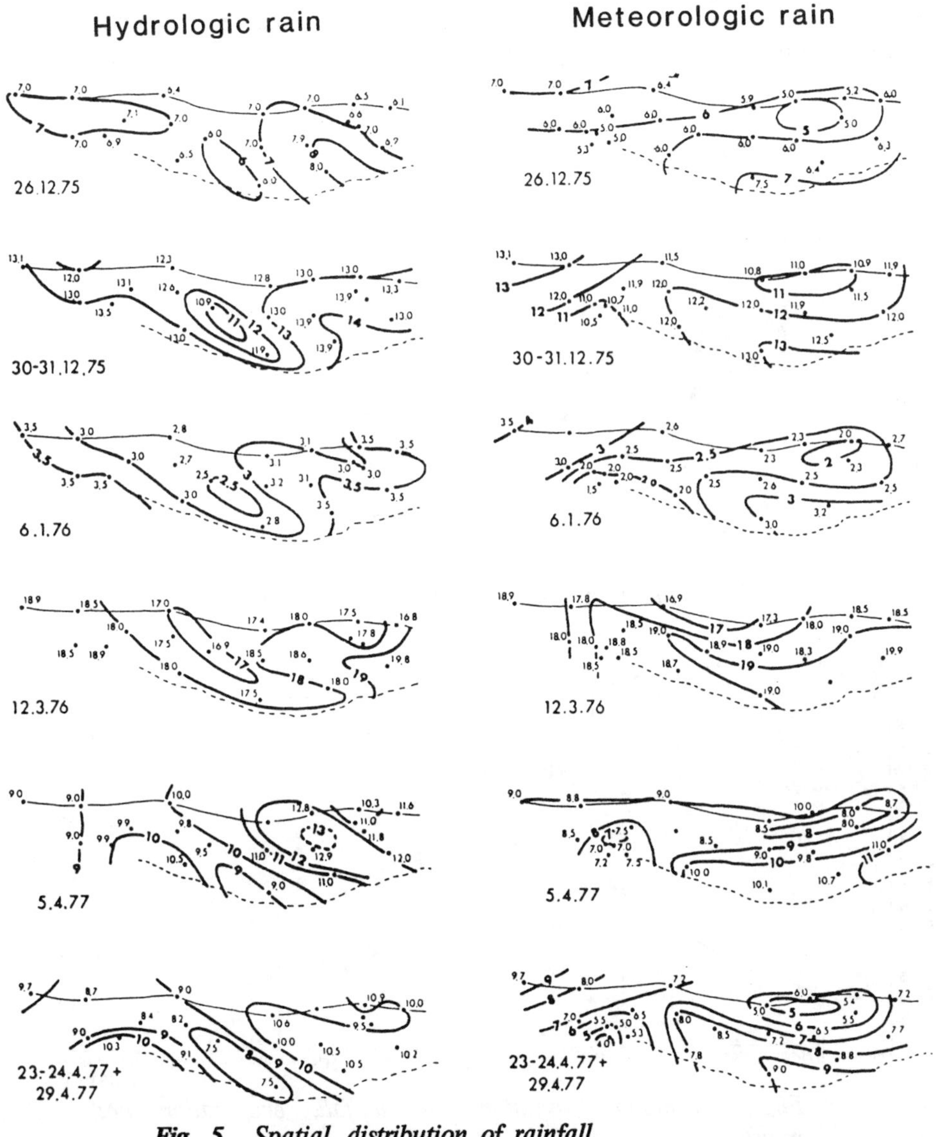

Fig. 5 Spatial distribution of rainfall.

speed gradient near the ground is greater. Between these two points we may assume that this ratio has intermediate values.

Infiltration capacity curves for each surface unit are shown in Fig. 6. An example of the results of evaporation and soil moisture measurements is shown in Fig. 7. The average potential evaporation in the summer is 7 mm day^{-1} and in winter it is 3 mm day^{-1}. Immediately after each rainstorm there is a sharp increase in soil moisture followed by a decline due to soil moisture losses. As evaporation during the short rainstorms is negligible, soil moisture losses from the upper 3 cm of the soil during these periods can be attributed to soil drainage. The soil moisture decay curves were used for the computation of soil moisture losses during the storm as a function of soil water content (Fig. 8). Soil moisture losses equal zero when soil water content is equal to zero, and soil moisture will not exceed a maximum value which represents field capacity (Lavee, 1982).

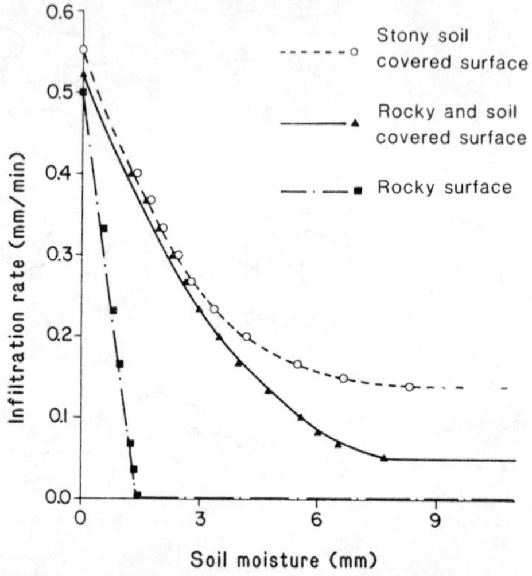

Fig. 6 Infiltration rates on different surface units.

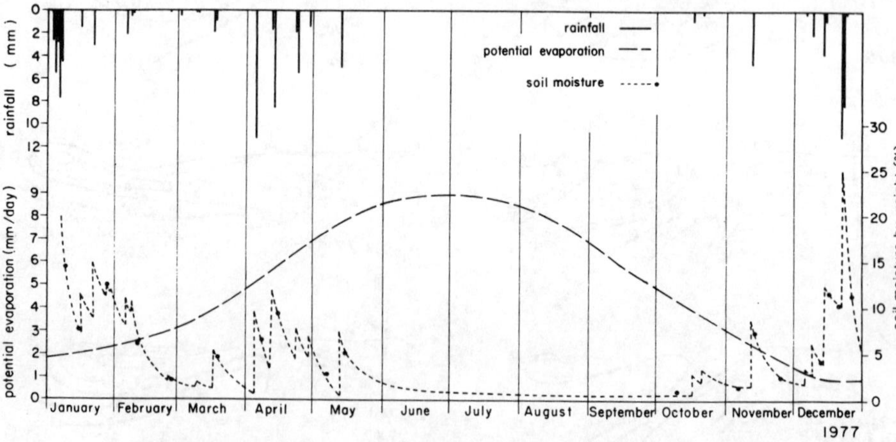

Fig. 7 Temporal variations of rainfall, evaporation and soil moisture.

Runoff response to rainfall is very quick (Fig. 9) and indicates that Hortonian overland flow occurs frequently in the study area. Overland flow data analysis (Table 1) shows that:

(a) Overland flow generation is spatially non-uniform due to pronounced differences in infiltration rates. Figure 10 for example, illustrates that the specific overland flow yield from plot 4C, which drains a rocky surface, is always greater than that of plot 4B, which drains a stony soil surface.

(b) The specific overland flow yield from the plots draining the whole slope (A) is lower than from the two adjacent short plots (B + C) combined (Fig. 11). This indicates significant infiltration within the hillslope before reaching the channel (Yair *et al.*, 1980). These results were obtained not only in plot 4, where infiltration losses increase downslope on passing from the rocky to the colluvial slope section (Fig. 11(a)), but also in plot 9 where the rocky section forms the lower part of the slope (Fig. 11(b)).

We may conclude, therefore, that overland flow discontinuity can be explained in terms of rainfall characteristics and the possible compounding effect of surface properties. Due to the short duration of rain showers, the infiltration rate at the end of the shower does not reach the minimum infiltration rate and

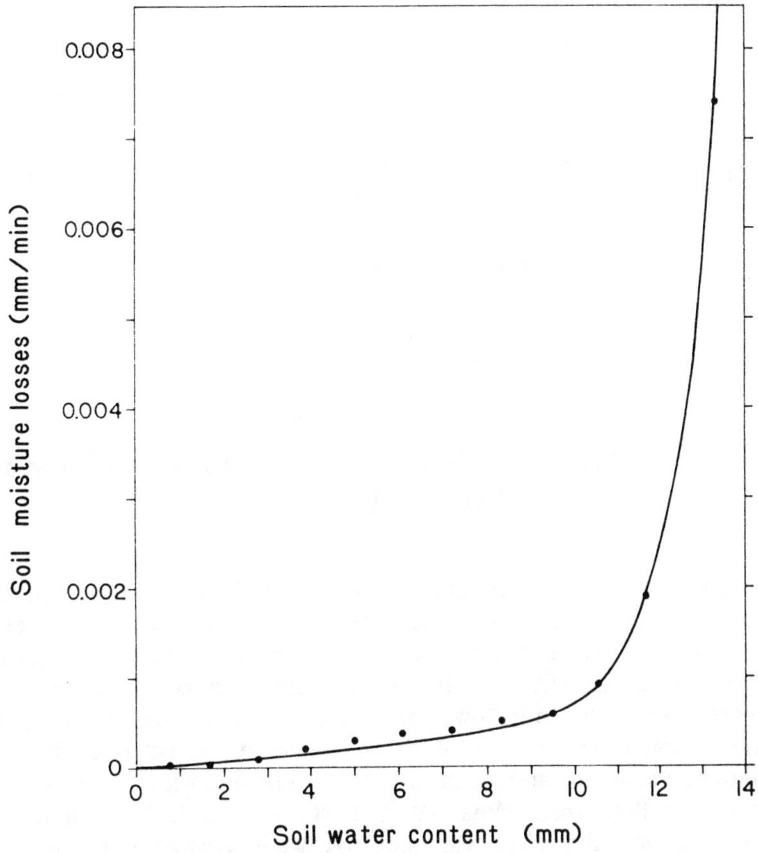

Fig. 8 Soil moisture losses rate.

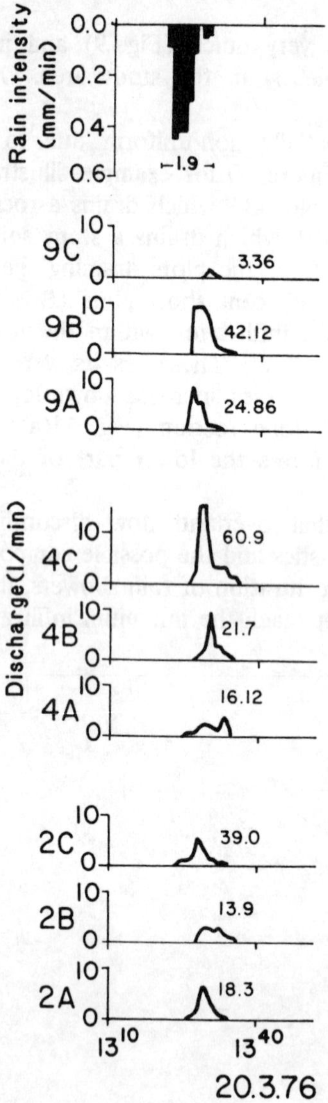

Fig. 9 Overland flow response to rainfall, 20 March 1976 (figures indicate runoff yield in l).

remains relatively high. Under such conditions the recession time is short, lasting 3 to 10 min. The combination of short rain showers, short recession time, and low flow velocities (1–3 m min^{-1}) cause the total duration of the flow to be shorter than the concentration time. In other words, the overland flow contributing area is limited to a belt close to the channel. The length of this belt can be expected to increase with rain shower duration and intensity. In a few instances plot 9A yielded more overland flow than plots 9B and 9C combined. This occurred when rain amount and intensity over plot 9A were greater than over plot 9B and 9C (Lavee, 1982).

***Table 1** Specific overland flow yield ($l\ m^{-1}$); * indicates no data*

DATE	4A.	4B	4C	4B+4C	9A.	9B	9C	9B+9C
11.11.75	*	0.020	0.799	0.098	0.209	*	0.000	*
26.12.75	0.506	*	*	*	0.589	0.789	0.554	0.691
26.12.75	0.013	0.000	*	*	0.051	0.075	0.030	0.056
26.12.75	0.000	0.000	*	*	0.024	0.025	0.032	0.028
30.12.75	0.000	0.000	*	*	0.032	0.071	0.000	0.041
30.12.75	0.000	0.000	*	*	0.000	0.000	0.000	0.000
30.12.75	0.171	0.179	*	*	0.193	0.254	0.363	0.300
30.12.75	0.000	0.000	*	*	0.011	0.019	0.000	0.011
31.12.75	0.000	0.000	*	*	0.011	0.023	0.000	0.013
06.01.76	0.000	0.000	0.117	0.043	0.000	0.033	0.000	0.019
03.02.76	0.000	0.000	0.094	0.035	0.011	0.067	0.000	0.039
23.02.76	0.000	0.000	0.000	0.000	0.090	0.229	0.000	0.133
24.02.76	0.000	0.000	0.031	0.011	0.000	0.020	0.000	0.012
25.02.76	0.000	0.000	0.016	0.006	0.000	0.020	0.000	0.012
12.03.76	0.000	0.000	0.126	0.047	0.014	0.094	0.031	0.070
12.03.76	0.571	0.460	1.554	0.867	0.604	0.947	0.844	0.904
12.03.76	0.195	0.232	0.561	0.355	0.299	0.397	0.274	0.346
12.03.76	0.000	0.000	0.095	0.035	0.027	0.042	0.000	0.024
12.03.76	0.097	0.108	0.424	0.225	0.247	0.344	0.167	0.270
12.03.76	0.000	0.000	0.022	0.008	0.012	0.024	0.000	0.014
20.03.76	0.027	0.064	0.311	0.156	0.074	0.232	0.026	0.146
12.04.76	0.000	0.000	0.034	0.012	0.000	0.030	0.000	0.018
13.04.76	0.000	0.000	0.021	0.008	0.000	0.061	0.000	0.035
18.05.76	0.000	0.000	0.038	0.014	0.000	*	0.031	*
24.10.76	0.000	0.000	0.249	0.092	0.099	0.244	0.000	0.142
04.01.77	0.126	0.291	0.750	0.462	0.234	0.506	0.135	0.351
04.01.77	0.000	0.000	0.018	0.007	0.000	0.000	0.000	0.000
05.01.77	0.000	0.000	0.143	0.053	0.012	0.000	0.000	0.000
06.01.77	0.000	0.019	0.357	0.144	0.090	0.187	0.000	0.109
06.01.77	0.000	0.000	0.000	0.000	0.000	0.030	0.000	0.018
06.01.77	0.104	0.237	0.490	0.332	0.156	0.525	0.245	0.408
06.01.77	0.000	0.000	0.061	0.023	0.000	0.039	0.000	0.022
06.01.77	0.000	0.000	0.117	0.043	0.000	0.039	0.000	0.022
07.01.77	0.000	0.000	0.020	0.008	0.000	0.000	0.000	0.000
07.01.77	0.025	0.053	0.511	0.223	0.075	0.143	0.168	0.154
22.01.77	0.000	0.000	0.015	0.006	0.000	0.017	0.000	0.010
22.01.77	*	0.040	0.250	0.092	0.096	0.110	0.077	0.096
06.02.77	0.000	0.000	0.054	0.020	0.000	0.035	0.000	0.020
17.03.77	0.000	0.000	0.000	0.000	0.000	0.031	0.000	0.012
05.04.77	*	2.330	4.937	3.213	3.495	5.442	4.592	5.086
13.04.77	0.061	0.030	1.863	0.708	0.312	1.046	0.702	0.902
24.04.77	0.116	0.163	0.836	0.413	0.286	0.587	0.462	0.535

DATE	4A.	4B	4C	4B+4C	9A.	9B	9C	9B+9C
24.04.77	0.000	0.000	0.016	0.006	0.000	0.025	0.000	0.015
24.04.77	0.000	0.000	0.023	0.009	0.000	0.033	0.000	0.019
13.05.77	0.812	0.739	1.292	0.948	0.680	1.045	1.168	1.097
12.11.77	0.008	0.000	0.295	0.109	0.099	0.215	0.085	0.162
12.11.77	0.000	0.000	0.203	0.075	0.047	0.100	0.000	0.058
09.12.77	0.000	0.000	0.116	0.043	0.021	0.077	0.000	0.045
14.12.77	0.034	0.093	0.315	0.177	0.154	0.286	0.116	0.215
14.12.77	0.000	0.000	0.105	0.039	0.047	0.154	0.034	0.104
22.12.77	0.863	1.517	1.706	1.595	1.265	1.240	1.234	1.240
22.12.77	0.406	0.600	0.699	0.640	0.710	*	0.448	*
22.12.77	0.156	0.309	0.482	0.388	0.529	*	0.160	*
22.12.77	0.281	0.407	0.424	0.415	0.411	*	0.338	*
23.12.77	0.000	0.070	0.187	0.114	0.064	*	0.086	*
23.12.77	0.010	0.087	0.177	0.120	0.024	0.000	0.084	0.035
23.12.77	0.000	0.028	0.123	0.063	0.010	0.020	0.000	0.012
23.12.77	0.000	0.011	0.213	0.086	0.000	0.000	0.000	0.000
23.12.77	0.218	0.264	*	*	0.056	0.000	0.116	0.048
23.12.77	0.000	0.000	0.017	0.006	0.000	0.000	0.000	0.000
30.03.78	0.015	0.062	0.301	0.151	0.111	0.221	0.054	0.151
24.04.78	0.010	0.024	0.220	0.095	0.111	0.160	0.046	0.112
15.10.78	0.032	0.052	*	*	0.088	0.187	0.070	0.138
12.12.78	0.142	0.243	0.577	0.369	0.156	0.451	0.268	0.375
12.12.78	0.953	0.911	1.241	1.038	1.102	0.858	1.401	1.085
12.12.78	0.800	0.944	1.072	0.997	0.812	0.611	1.018	0.781
12.12.78	1.292	1.149	1.843	1.412	1.548	0.820	1.424	1.072
12.12.78	0.066	0.145	0.235	0.179	0.228	0.127	0.122	0.125
12.12.78	0.280	0.309	0.505	0.383	0.241	0.083	0.253	0.154
08.01.79	0.007	0.039	*	*	0.081	0.177	0.030	0.116
08.01.79	0.099	0.129	*	*	0.201	0.275	0.230	0.256
08.01.79	0.173	0.215	*	*	0.119	0.086	0.310	0.179
09.01.79	0.049	0.101	*	*	0.104	0.074	0.308	0.171
09.01.79	0.674	0.834	*	*	0.838	0.820	2.277	1.429
09.01.79	0.000	0.000	*	*	0.017	*	0.000	*
09.01.79	0.843	1.671	*	*	1.261	0.912	0.815	0.871
22.01.79	0.245	0.289	0.276	0.285	0.085	0.138	0.000	0.080
23.01.79	0.000	0.000	0.077	0.028	0.000	0.060	0.000	0.033
23.01.79	*	0.416	0.546	0.466	0.335	0.509	0.589	0.542
07.02.79	0.000	0.000	0.000	0.000	0.016	0.114	0.000	0.066
09.02.79	0.060	0.350	0.771	0.508	0.242	0.513	0.401	0.466
09.03.79	0.000	0.000	0.000	0.000	0.013	0.138	0.000	0.080
09.03.79	0.000	0.000	0.000	0.000	0.043	0.118	0.059	0.093
04.05.79	0.137	*	0.469	*	0.130	0.176	*	*

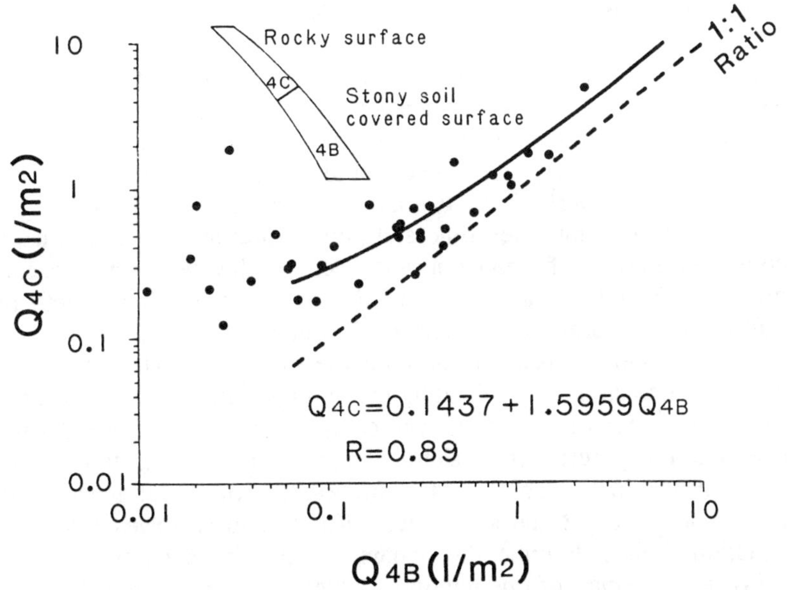

***Fig. 10** Specific overland flow yield: the effect of surface properties.*

COMPUTER SIMULATION OF OVERLAND FLOW

Given the above, it is clear that overland flow in the study area is very dynamic both in time and space due to two main factors; the great temporal and spatial variability of the rainfall intensity and infiltration rate, and the typically short duration of rain showers and resulting limited rainfall.

Given this complex situation, statistical rainfall-runoff relationships, which are based on average values for large areas and long time units, do not describe the actual field condition. In order to show the influence of the spatial and temporal variability of hydrological rainfall as well as that of the infiltration rate on the spatial and temporal variability of overland flow generation and its continuity along a slope, a computerized deterministic simulation was developed (Lavee, 1986).

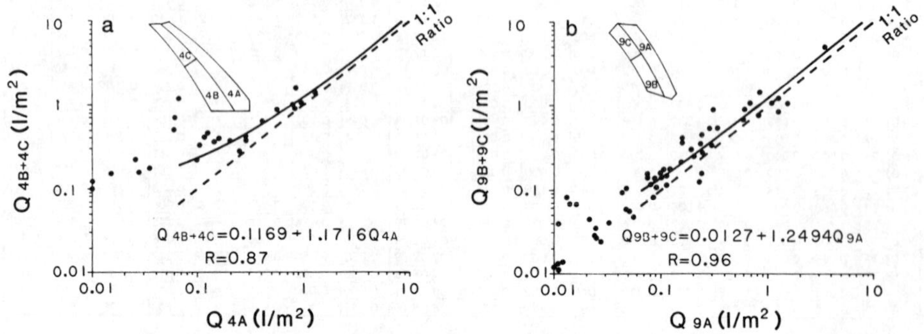

Fig. 11 Specific overland flow yield: the effect of rainfall properties.

DESCRIPTION OF THE SIMULATION

The computer program is written in version 360 of the Continuous System Modelling Program (CSMP) language (IBM, 1969; Hillel, 1977). The model provides a quantitative description of the generation and continuity of overland flow as a function of specific rain conditions and surface properties. It is based on the Hortonian approach to overland flow generation. Therefore the depth of overland flow at any given time and point represents the difference between the combined amount of direct rain and overland flow which had reached the point up to a given time, and the quantity of water which had left the point by infiltration, evaporation and overland flow up to that time.

The model takes into account the spatial variability of surface properties and rain characteristics by dividing the hillslope into *N* compartments, all parallel to the slope contours. Within each compartment, the processes associated with overland flow generation are simulated, i.e. rainfall, infiltration, evaporation and soil moisture losses. Though these processes occur simultaneously, they do not necessarily occur at the same rates. The computer solves a set of differential equations which describe the dynamic system by computing numerically over short time intervals of one minute. At the end of each interval, the overall effect is summed and the surface water from each compartment "flows" to the adjacent compartment further down the slope. Each variable is then updated and reset for the beginning of the next time interval.

THE VARIABLES

The model variables are shown in Fig. 12.

Hydrological rainfall intensity (*RAIN*) in each time interval and compartment is calculated by:

$$RAIN\ (1,N) = RAINR \times RRR\ (1,N) \times RR\ (1,N)$$

where *RAINR* = rainfall intensity measured by a recorder; *RRR* = ratio between rainfall amounts in tilted gauge and recorder; and *RR* = ratio between rainfall amounts at 30 cm and at ground level.

The **potential evaporation rate** *(EVAP)* and the **infiltration capacity** (*INF*) both vary in time and over space. Data entered to the program are based on field measurements.

Actual evaporation (*RVAP*) and **actual infiltration** (*RF*) for each compartment and time interval are calculated as follows:
(a) If $TRUN + RAIN \geqslant INF + EVAP$, then $RF = INF$ and $RVAP = EVAP$;
(b) If $INF + EVAP > TRUN + RAIN > INF$, then $RF = INF$ and $RVAP = TRUN + RAIN - INF$;
(c) If $TRUN + RAIN \leqslant INF$, then $RF = TRUN + RAIN$ and $RVAP = 0$;
where *TRUN* = overland flow depth at the beginning of the time interval.

Soil moisture (*STOR*) in each compartment has an initial value (*ISTO*) at the beginning of each storm. Soil moisture increases by actual infiltration and decreases with actual soil moisture losses through evaporation from the soil and by seepage.

The **potential soil moisture losses rate** (*LOSS*) in each compartment varies with time as a function of soil water content (Fig. 8). **Actual soil moisture losses** (*SLOS*) are calculated as follows:
(a) If $STOR < FCAP$, then $SLOS = LOSS$;
(b) If $STOR \geqslant$ FCAP and $RF \geqslant$ LOSS, then $SLOS = RF$ (soil moisture does not change);
(c) If $STOR \geqslant FCAP$ and $RF < LOSS$, then $SLOS = LOSS$ (soil moisture decreases);
where *FCAP* = field capacity.

Depth of overland flow (*RUN*) on each compartment equals zero at the beginning of the simulation. It changes in each time interval according to the above-mentioned processes. At the end of each time interval, the water is moved lower down the slope to the adjacent compartment.

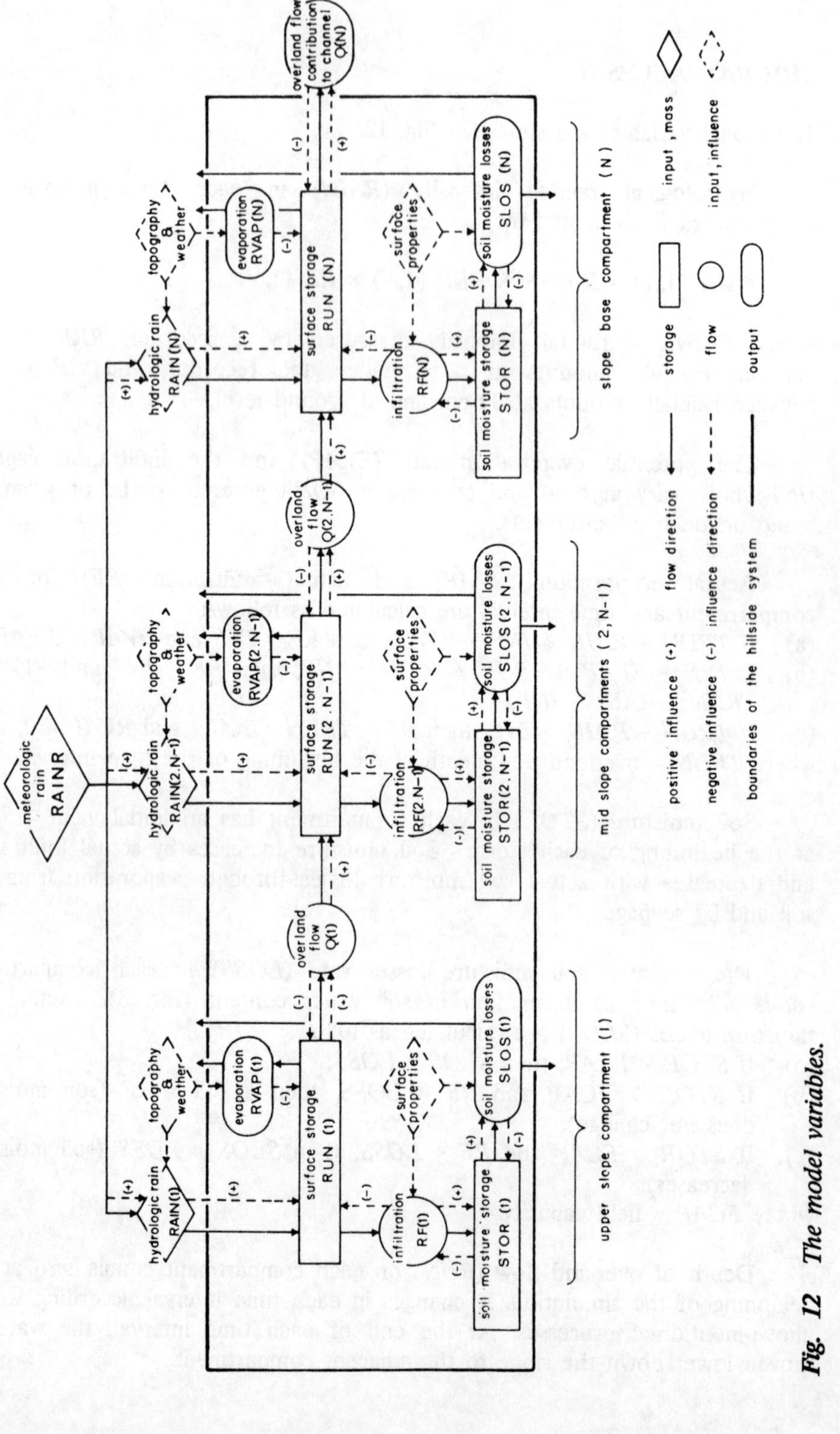

Fig. 12 The model variables.

APPLICATION TO OVERLAND FLOW CONTINUITY

The simulation program was verified by comparing simulated hydrographs with the actual ones (Fig. 13). The results show that the model fits well with the actual field situation. The model was then used to check the effect of short rain showers on overland flow continuity along slopes having uniform surface properties. The program was operated on six hypothetical plots. The infiltration capacity over the whole plot area was chosen to be equivalent to that of the stony soil cover surface; thus surface property differences do not

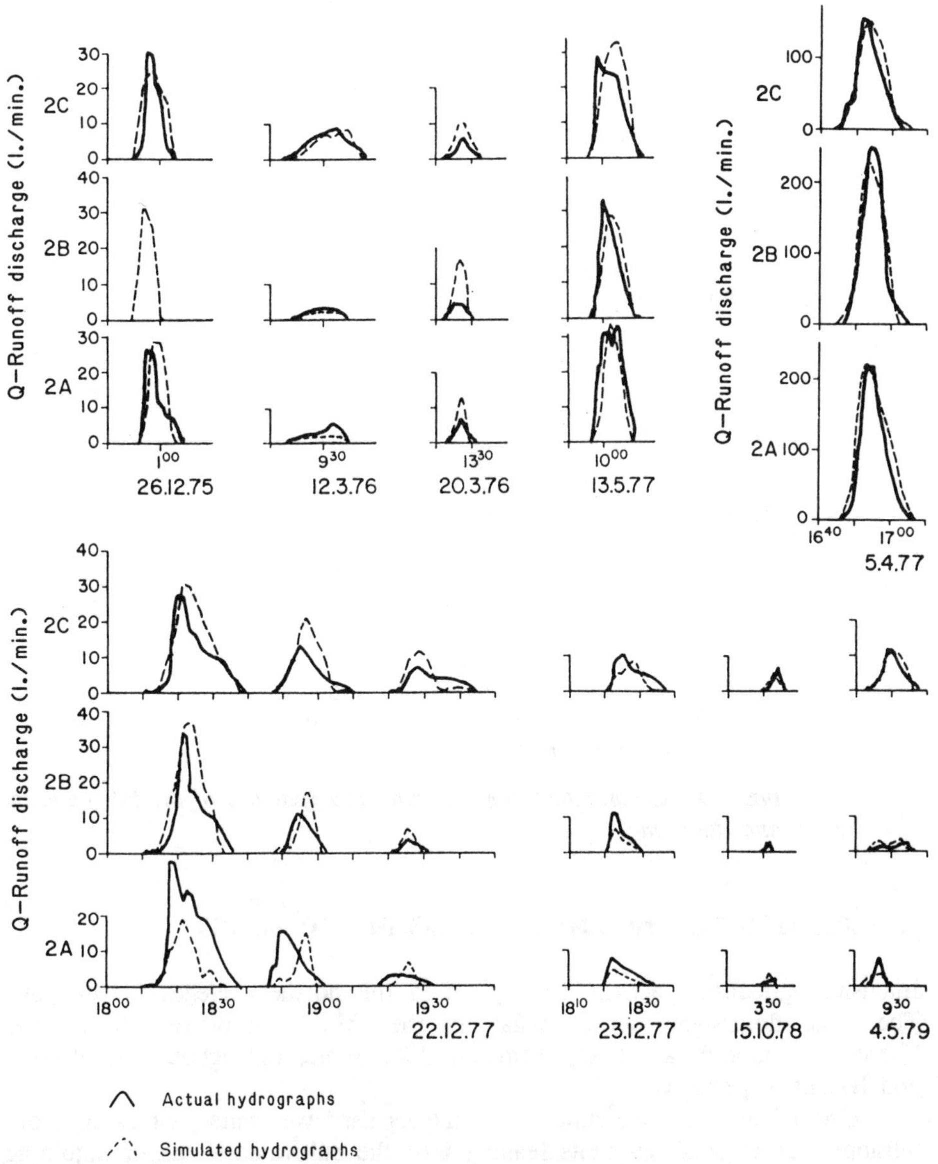

Fig. 13 Actual and simulated hydrographs.

affect the continuity of the flow. The only difference between the plots was in their respective lengths - 15, 30, 45, 60, 75 and 90 m. Sixteen rain showers were simulated on each plot for four different durations and for four different intensities (Fig. 14). The intensities and durations were chosen on the basis of an analysis of the rainfall properties at the experimental site. The relationship between runoff yields from the six plots and their lengths identifies the length of the slope contributing overland flow to the channel. Figure 14 indicates that for the above-mentioned surface properties, with a rain shower lasting 30 min with an average intensity of 12 mm hour^{-1}, the length of the slope contributing overland flow to the channel is 30 m. An increase in the contributing slope length is expected with increasing rain shower duration and/or intensity.

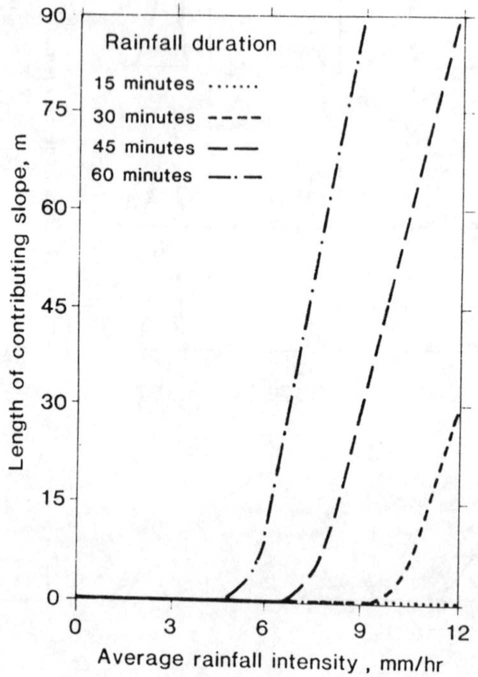

Fig. 14 Contributing slope length as a function of rainfall intensity and duration.

APPLICATION TO ANCIENT FARMING IN THE NEGEV

Intensive agricultural activity took place in the northern Negev in the past. The most flourishing period began in the third century BC during the Nabatean era, and lasted approximately 1000 years, throughout the Roman and Byzantine periods.

One of the main agricultural techniques used was water harvesting from hillslopes, using small conduits leading into the cultivated fields of adjoining valleys. These conduits can be seen today (Fig. 15) crossing the contour lines

Fig. 15 Overland flow collecting conduits.

at an acute angle. Parallel conduits, sometimes as many as 10 on one slope, can be seen. The distance between two adjoining conduits is usually 10 to 15 m.

The distribution of the conduits raises a question. Why are they so numerous on the slopes and why were the ancient farmers not satisfied with one conduit at the base of the slope? Evenari *et al.* (1982, p. 109-110) have dealt with this question and state "in this way the overall runoff was divided into small streams of water so that large flash floods were prevented."

A complementary explanation can be provided by the hydrological results presented in this paper. Under typical short rain shower conditions, which prevail in the study area, a continuous overland flow usually occurs for slope lengths of 10 to 15 m. Therefore the spatial distribution of the conduits provides the most efficient system for water harvesting.

Apparently the ancient farmers were well aware of the phenomenon of overland flow discontinuity. Through observation, they may have come to realize that the overland flow generated at the upper part of the slopes had little chance of reaching the valley and would be lost without conduits to collect the overland flow on its way downslope. They also appear to have recognized that a shorter separation between conduits would enable them to receive water even for lower-magnitude rainfall events when flow distance is relatively limited.

REFERENCES

Evenari, M., Shanan, L. & Tradmor, N. (1982) *The Negev.* (2nd Edn) Harvard University Press, Cambridge, Massachusetts, USA.

Hillel, D. (1977) *Computer Simulation of Soil-Water Dynamics.* International Development Research Centre, Ottawa, Canada.

Horton, R. E. (1945) Erosional development of streams and their drainage basin. *Bull. Geol. Geol. Soc. Am.* **56**, 275-370.

IBM (1969) *Continuous System Modelling Program, User's Manual.* Program Number 360a-cx-16x, IBM Corp. Technical Publications Dept., New York USA.

Lavee, H. (1982) Distribution of slope areas that contribute runoff to the stream in an arid environment. PhD Thesis, Hebrew Univ. of Jerusalem, Israel (in Hebrew).

Lavee, H. (1986) A deterministic simulation model for rainfall-runoff relationship on arid slopes. *Z. Geomorphol. Suppl. Bd* **58**, 35-46.

Morin, J., Cluff, B. C. & Powers, W. R. (1970) Realistic rainfall simulation for field investigations. Paper 78, *51st Annual Meeting American Geophysical Union,* Washington, DC., USA.

Sharon, D. (1980) The distribution of hydrologically effective rainfall incident on sloping ground. *J. Hydrol.* **46**, 165-188.

Yair, A., Sharon, D. & Lavee, H. (1978) An instrumented watershed for the study of partial area contribution of runoff in the arid zone. *Z. Geomorphol. Suppl. Bd* **29**, 71-82.

Yair, A., Sharon, D. & Lavee, H. (1980) Trends in runoff and erosion processes over an arid limestone hillside, Northern Negev, Israel. *Hydrol. Sci. Bull.* **25**, 243-255.

Erosion, Transport and Deposition Processes (Proceedings of the Jerusalem Workshop, March-April 1987). IAHS Publ. no. 189, 1990.

Towards a dynamic model of gully growth

ANNE C. KEMP née MARCHINGTON
*Department of Geography, University of Bristol, University Road, Clifton, Bristol BS8 1SS, UK**

Abstract Headward migration of gullies around the margins of drainage networks may be a major factor in the increased sediment yields observed in semi-arid southeast Spain. Although many processes have been cited, the precise nature of this gully extension and bifurcation is not known. A gully which is migrating into a hillslope may be considered by way of its morphological dynamics. A digital model has been formulated in which the governing process is erosion by overland flow. Using finite techniques, the continuous phenomena may be approximated by discrete functions. The model grid is set up by defining strip catchments which are bounded by orthogonal flowlines. Overland flow is then routed downslope kinematically, assuming gravity and friction as the controlling forces. Sediment detachment, transport, and deposition are estimated using a Musgrave-type approach which incorporates an interaction term. This compares energy required to carry sediment already in transport with the total capacity of the flow to do work. The consequent change in the slope surface is expressed by migration of the contours along the flowlines. In this manner, the effects of hillslope and gully geometry on gully development may be explored. The simulations indicate a critical balance between a linear propagation of the erosion headwards, and a diffusion laterally of this impulse. Within this balance is identified a possible mechanism for the bifurcation phenomenon. These results are being verified by laboratory experimentation.

INTRODUCTION

This paper reports an investigation into the possible mechanisms of gully head migration. More specifically, the role of gully head geometry and slope morphology in controlling the processes which lead to the extension and bifurcation of gullies is explored.

The objectives of this paper are to present:

a) a theoretical model of gully growth and bifurcation which offers a simple and easily formulated framework on which to base experimental studies; and
b) a two dimensional digital model which is being developed to simulate the dynamics of soil erosion, and the resultant change in the

---

* Now at Barnus Ltd, Environmental Engineering and Consultancy Services, Thorncroft Manor, Dorking Road, Leatherhead, Surrey KT22 8JB, UK.

morphology of the hillslope and gully head.

The growth of rills and their development into gullies and even badlands erosion increase by orders of magnitude once rilling and gullying ensue, and any model which fails to take account of their development will be of limited value in many conservation situations. High sediment yields are often attributed to surface wash, yet evidence in southeast Spain indicates that the problem actually originates from the headward extension of gullies around the drainage net margins (Thornes, 1976, p.41, p.70; Thornes & Gilman, 1983, p.133).

MODELLING GULLY GROWTH

Sheet erosion can be directly related to the bottom shear stress produced by overland flow. Field evidence (Dunne, 1980; Dunne & Aubrey, 1986) and laboratory evidence (Moss *et al.*, 1982) indicate that sheet flow is inherently unstable and will split into small concentrated rivulets of flow. Rainfall, micro-topography and vegetation will have a considerable effect on this tendency.

In a general sense, the initiation and growth of rills and gullies is dependent on a sufficient concentration of this bottom shear stress to form a definable channel. Surrogates for the amount of discharge, such as the length or area of contributing slope, have been discussed by Horton (1945) and Schumm (1956), respectively. Schumm proposes a constant of channel maintenance, essentially the drainage area required to support a given length of channel. The underlying implications are of great significance for channel extension. In the case where full extension of a channel has been reached, such as in Fig. 1 for a hypothetical planar surface, if all other factors remain constant further growth can only occur if a length or area is obtained greater than already existing. A curvature of the contours must occur which is sufficient to induce a greater than critical component of length or area.

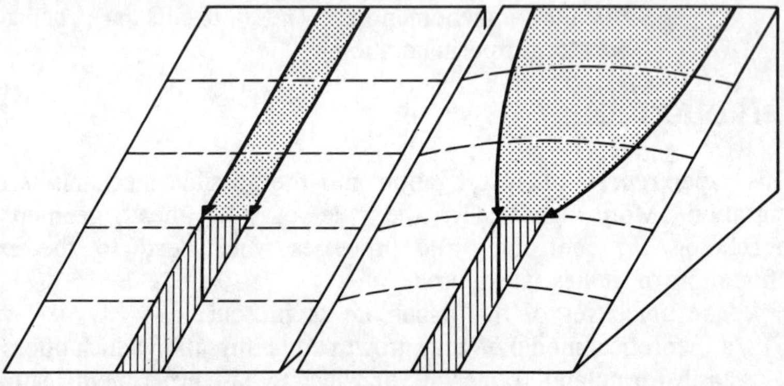

***Fig. 1** Drainage area required for channel maintenance: where full extension of a channel has occurred (a), growth may be re-initiated if the contributing area is increased above a critical component by a curvature of the contours (b).*

Small depressions or nicks may be initiated and enlarged as a result of some sort of perturbation, such as a local variation in vegetation, surface roughness or surface crusting. The growth of a hollow will increase the curvature of contours across a hillslope, and will therefore lead to a convergence of water and sediment. The hollow will grow if the increase in the rate of work is greater than the increase in sediment to be transported. Smith & Bretherton (1972) and Kirkby (1980a) formalize these relationships using the following instability criterion:

$$A \frac{dS}{dA} > S \tag{1}$$

where A is contributing area
S is sediment in transport.

If the increase in the amount of sediment to be transported (right hand side of the equation) is greater than the increase in capacity to transport (left hand side), infilling of the hollow will occur. Growth can only occur if there is a relative increase in the transporting power.

Clearly, the influence of three-dimensional topography is of the greatest importance to drainage net development and reflects the self-generating nature of drainage channels. Where there is a flow convergence and an adequate concentration of erosional power, channel growth is possible.

Rills and gullies are dynamically similar to channels, but are characterized by ephemeral flow, and a close coupling to the hillslope. Within active gullies, near-vertical scarps can develop at the head of the channel. Once a headcut is initiated, it may retreat upslope into otherwise undisturbed hillslopes. Many examples have been observed in southeast Spain. Channel storage will be reduced where there is a concentration of water in the gully, and erosive power will increase in relation to the flow depth. Runoff over the headcut may contribute to gully growth by exerting stresses on the channel boundary, by removing accumulated soil debris from the channel and by eroding the gully banks through undercutting (Francis, 1985).

BIFURCATION OF GULLIES

The question then arises as to why a headward-extending gully should branch. Thornes (1984) has proposed an analytical model in which the geometry of the gully head controls the branching, or bifurcation, phenomenon. Once an initial perturbation has developed, in this context perceived as a headcut, or slope failure boundary, an erosional pulse or signal will pass up throughout the system. Its forward and lateral velocities will vary as a function of the imbalance between force and resistance.

Assume that the form of the pulse is analogous to a shock wave migrating through the hillslope. It is propagated linearly along the line of greatest erosive power, calculated here per unit width of bed slope at the shock boundary. Overland flow, taken as the generator of this erosive power, is assumed to be orthogonal to the contours. The contours themselves will migrate as erosion proceeds. The lateral dissipation of the erosional energy is controlled by the relative strength and cohesiveness of the bank material.

If the gully head is in a fixed location, it is assumed that a wave of erosion propagates outwards with a velocity V_s. This is mainly determined by processes such as creep and mass failure which are controlled by the properties of the materials. If migration is wash-controlled, then it is also moving upslope with a propagational velocity, V_x, as shown in Fig. 2. Where the ratio of the velocities is greater than unity:

$$\frac{V_x}{V_s} > 1 \qquad (2)$$

then the shock travels ahead of the outwardly propagated waves, and the opportunity time for relative widening of the channel is less. The shock is 'supersonic'. If this criterion does not hold, the shock is 'subsonic', and growth is influenced by the outward movement of the propagating wave. In semi-arid environments the wash (forward) component occurs intermittently, and other processes, such as weathering, creep and local wash and splash, may diffuse the gully head outwards. This will be particularly the case on weak lithologies or on surfaces which do not generate much overland flow.

It is suggested that the geometry of a gully head governs the distribution

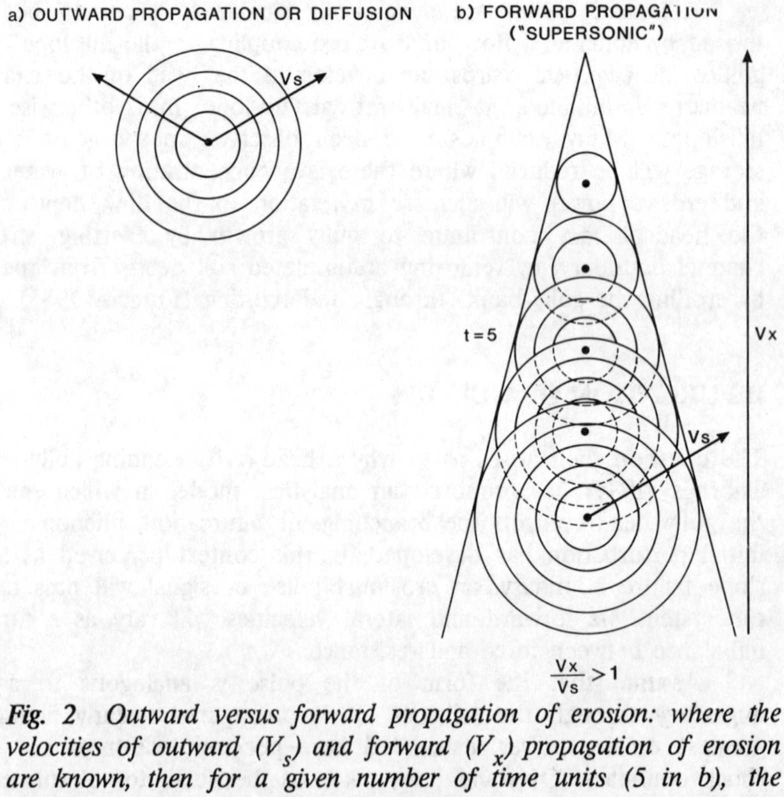

Fig. 2 Outward versus forward propagation of erosion: where the velocities of outward (V_s) and forward (V_x) propagation of erosion are known, then for a given number of time units (5 in b), the resulting shape of the gully may be predicted.

of shear stress around its boundary, through its spatial relation with the orthogonal flowlines from upslope. In Fig. 3, it is apparent that, as one moves around the gully head boundary away from the apex, or central axis of the gully head, the angle at which the flowlines intersect the gully boundary decreases. Each unit width of flow from upslope has to cross a progressively larger width of gully boundary. Thus the concentration of erosive power on the boundary decreases. If this concentration of erosive power is greatest at the apex, extension will continue linearly. If deflected to either side, then branching of the gully may occur.

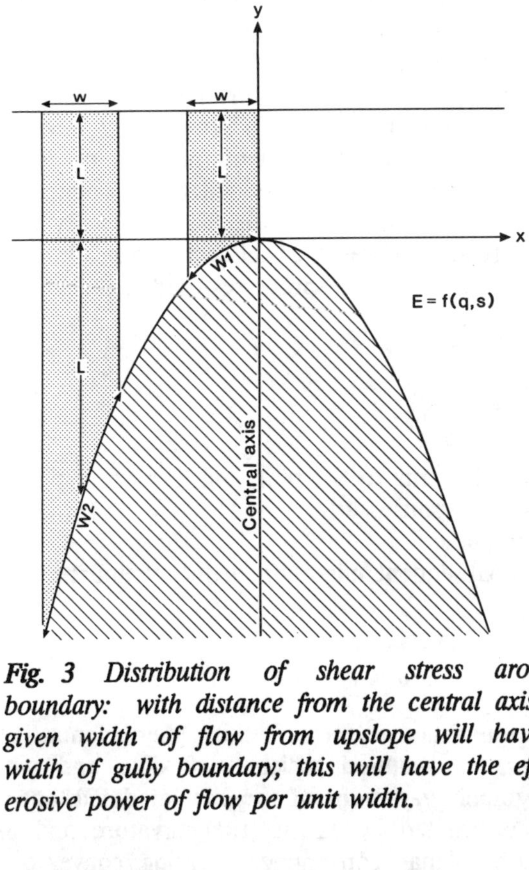

Fig. 3 Distribution of shear stress around a gully head boundary: with distance from the central axis of the gully head, a given width of flow from upslope will have to cross a greater width of gully boundary; this will have the effect of decreasing the erosive power of flow per unit width.

Thornes (1984) suggests that the relative magnitudes of forward and lateral migration of the gully walls may be a critical control of bifurcation. In southeast Spain, low density drainage nets may be observed in lithologies with a high shear strength. Supersonic migration should lead to the formation of long narrow channels, due to the limited opportunity for branching. Where subsonic conditions exist, the widening of channels by active slumping and other processes, such as creep, may increase the likelihood of branching. Higher density dendritic networks should develop. Thornes' model describes the conditions of hillslope and gully head geometry, in which bifurcation is likely to occur.

DIGITAL SIMULATION

Once the point of bifurcation is reached, the analytical problem becomes intractable, and an alternative lies in numerical solution through digital simulation. This is a powerful tool for exploring some of the implications of such a conceptual model, in that the model system may be completely controlled. The model which is devised needs to be simple enough to be easily manipulated and understood, whilst being sufficiently representative of the natural system under scrutiny to provide a meaningful evaluation.

The key questions to be addressed are:

(a) how will the form of the gully change through time and space; and

(b) under what circumstances will the gully branch?

The modelling of overland flow, in the context of gully growth, is relatively straightforward, and is therefore an appropriate process to adopt in the initial assessment of this concept. To study gully development in this paper, a hydrodynamic model and a sediment transport model are coupled in order to account for the hydraulics of overland flow, sediment transport, and morphological considerations (cf. Cordova *et al.*, 1983). By routing sediment across the slope, it is then possible to calculate the migration of the contours and gully head. The dynamics of the processes, and the changing morphology may thereby be simulated.

PROGRAM OUTLINE

There are three major elements to the program:

(a) the morphological response surface;

(b) the overland flow; and

(c) the erosion, and subsequent migration of the contours.

Morphology

In developing a numerical model of this phenomenon, a generalized geometrical framework is required which is flexible and which can take account of the topographic variations of the simulated hillslope. Hypothetical initial surfaces are constructed by varying the curvature and interval of the contours. Slopes may be planar, converging, diverging, convex or concave. The two-dimensional hillslope is represented (Fig. 4) by an assembly of one-dimensional strip catchments which are assumed to be independent of each other. The strips are defined on the basis that the water flows in directions orthogonal to the topographic contours, the steepest route available. Within each strip, a number of cells are defined by the intersection of the orthogonal flow lines and contours. The average length, width, slope and area are calculated for each cell, and a roughness parameter is established. These cells are the basis for the numerical solution of the erosive process operating over the hillslope.

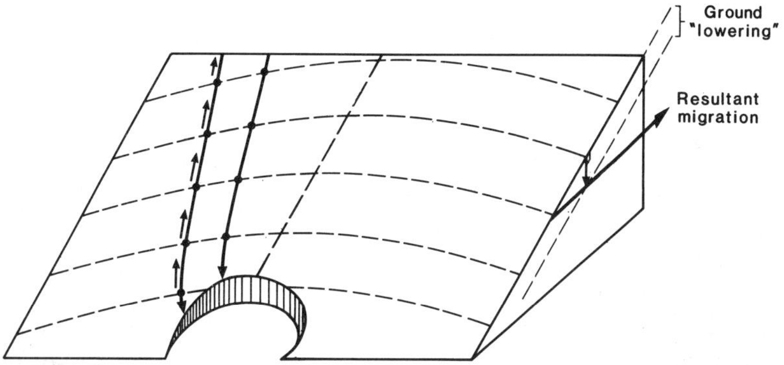

Fig. 4 Representation of two dimensional slope by an assembly of one dimensional strip catchments: strip catchments are defined by flow lines orthogonal to the contours, and migration of the contours is calculated along the flowlines, as a function of the perpendicular ground lowering.

Overland flow

The erosion-deposition algorithm is a second stage component of a runoff generating model. In the kinematic approximation of the hydrodynamic equations, which is used here, the controlling forces are the gravity and the friction terms. The assumptions of this approximation are that the slope is between 2° and 25° and is varied gradually. The final form of the equation, after Kibler & Woolhiser (1970), is derived from a combination of the equations of continuity and momentum:

$$\frac{\partial A}{\partial t} + \frac{\partial Q}{\partial x} = q \qquad (3)$$

where $A = w\,h$ is area;
w is width;
h is depth;
$Q = A\,u$ is discharge;
u is velocity;
q is input per unit area;
t is time; and
x is distance.

A stage equation is adopted to eliminate one of the two dependent variables:

$$u = n\,h^{m-1} \qquad (4)$$

where n is a roughness and slope coefficient;
m is an exponent.

This is substituted into equation (1) and the terms expanded, thus:

$$\frac{\partial w h}{\partial t} + \frac{\partial w n h^m}{\partial x} = q \tag{5}$$

The model assumes that overland flow occurs as turbulent sheet flow.

A subroutine based on the method of characteristics has been implemented for solution of the kinematic approximation. This allows a more accurate simulation of the hydrograph than has been achieved by finite difference methods, and it calculates the necessary time increment for stability. The equations are solved in a sequential manner, such that the dependent variable is determined for each node of the solution mesh. This is achieved by solving along the characteristics, and extrapolating back to the specified nodes. This numerical procedure will be described in greater detail in a later paper.

Formulation of the erosion algorithm

The modelling of overland flow generation across a hillslope is well-established in the literature. The formulation of an erosional-depositional algorithm is more problematic.

Ellison (1946) cites four critical features of the entrainment-transport situation:

(a) the detaching capacity of the erosive agent;
(b) the transport capacity of the erosive agent;
(c) the detachability of the soil; and
(d) the transportability of the soil.

Flow at a point in time and space has a given energy available for detaching and transporting soil. Foster & Meyer (1975) propose that the ratio of the sediment load to the transport capacity, a relative term of energy for transport, plus the ratio of the detachment rate to detachment capacity, a relative term of energy for detachment, equals unity, the total available energy, i.e.

$$\frac{D_f}{D_c} + \frac{G_f}{T_c} = 1 \tag{6}$$

where D_c is the detachment capacity;
D_f is the actual detachment rate;
T_c is the transport capacity; and
G_f is the sediment in transport.

The detaching and transporting capacity of the flow varies between inter-rill and rill areas. The dynamic nature of rilling is a difficult feature to model, not least because rainfall affects the presence of small scale channelized flow (Dunne & Aubrey, 1986). It is assumed firstly, that the erosive agent is overland flow, and that raindrop effects can be ignored, and secondly, that the estimation of detaching and transporting capacity can be assessed across the general region of the hillslope, for each cell of the spatial mesh.

The basis for most existing erosion algorithms was established by Zingg & Musgrave in the 1940s (Zingg, 1940; Musgrave, 1947) in which erosion rates are a function of surrogates for the shear stress of overland flow:

$$Y = Q^{1.66} S^{1.5} \tag{7}$$

where Y is sediment yield in $cm^3\ cm^{-1}\ year^{-1}$;
Q is water discharge; and
S is slope angle.

In the model, this equation is used to estimate the transporting and detaching capacities. The actual transport and detachment rates are determined by the interaction between transport and detaching capacities. The transport deficit approach discussed by Kirkby (1980a), is adopted so that the actual erosional rate approaches the detachment capacity when the sediment load is very much smaller than the transport capacity:

$$D_f = \frac{T_c - G_f}{T_c / D_c} \tag{8}$$

where: $T_c = a\ Q^b\ S^c$ (9)

$$D_c = k\ Q^m\ S^n \tag{10}$$

a, k are coefficients dependent on soil resistance; and
b, c, m, n are exponents.

The greater the difference between the transporting capacity and the detaching capacity, the slower the rate of uptake. As a result, there is a gradual rather than sudden transition from detachment-limited to transport-limited removal. This inter-relationship between detachment and sediment load explains the changes in the sediment yield even when other variables such as depth and energy grade-line remain constant.

The erosion-deposition algorithm uses the discharge values of each cell generated by the kinematic overland flow routine, the slope and soil parameters and the sediment concentration of the current time step. This enables sediment to be routed down the slope. The net change in surface throughout the storm is stored for every node on the surface.

At the end of each storm event, the change in topography is assessed. Ground lowering is reflected, not by changing the height values of each node, but by calculating the resultant migration of each node along the orthogonal flowline. This is illustrated in Fig. 4. By plotting the new positions of the contours, the change in the morphology is easily perceived through the deformation of the contour lines. Their relative velocities will alter the steepness of the slope and deformation will change the direction of the flow lines.

SIMULATIONS

The slope profile has an important effect on the distribution of erosion across the surface. The output of three runs is given in Fig. 5, where the three-dimensional plots show the total amount of ground lowering at a point, across the entire surface. For all these surfaces, overall erosion rates increase downslope, but the trends are significantly different. Here, the slope profile is respectively convex, planar, and concave. The average gradient is 6.3°, and the slope length is 2 m. On the convex slope, discharge and slope are increasing in a downslope direction, leading to a progressive increase in the total amount of ground lowering. On the planar slope, discharge only is increasing downslope. Once the increase in the transporting capacity no longer compensates the additional sediment to transport, the increase in ground lowering downslope declines. On the concave slope, there is an initially rapid increase in erosion downslope, but this levels out as the slope declines, and at the base of the slope, the rate of ground lowering begins to decrease. On a longer slope, deposition may occur, and a scanning routine is necessary to detect reverse slopes.

If it is assumed that the migration of the gully head is related to the erosive power of the flow, the position of the gully head in relation to the above trends becomes important. In the second set of simulations (Fig. 6), the same shape of gully head is considered at two different

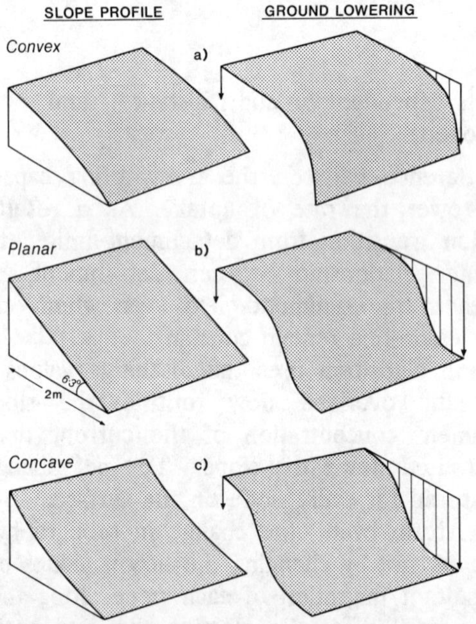

Fig. 5 Three dimensional plots of ground lowering across the entire hillslope at the end of the storm; the slope morphology is shown on the left; actual ground lowering, shown on the right, is increasing downslope.

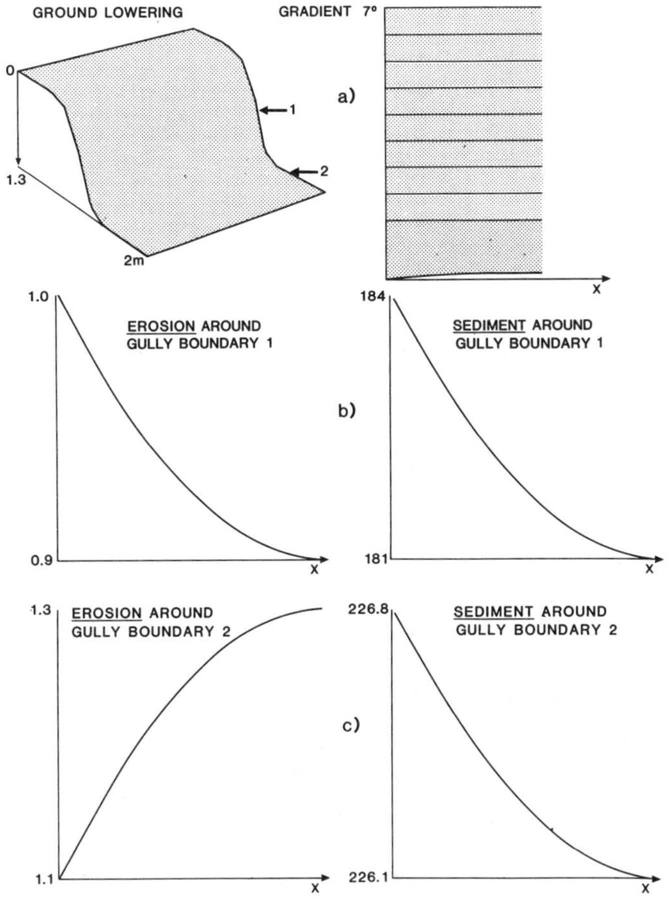

Fig. 6 *Gully head migration from two different positions on a slope: (a) the distribution of erosion (left), the plan form of the slope (right); (b) amount of erosion (left) and sediment yield (right) around the gully head for position 1 on slope; and (c) amount of erosion (left) and sediment yield (right) around the gully head for position 2 on slope.*

positions on a planar slope; firstly where the erosion rate is still increasing in the downslope direction; secondly where it is not. In the first case, the erosive power of the water increases as it flows downslope, over and above the increase in sediment load, and subsequently the curvature of the bottom boundary is obliterated. In the second case, the water's erosive power is no longer increasing downslope, and an increasing proportion is expended on transporting sediment. As a consequence, there is a concentration of erosion at the apex of the gully and it migrates forward with time.

In the final set of simulations presented here (Fig. 7), two different gully head shapes are considered on identical planar slopes, at a position where the erosion rates are still increasing downslope. In the first case, the

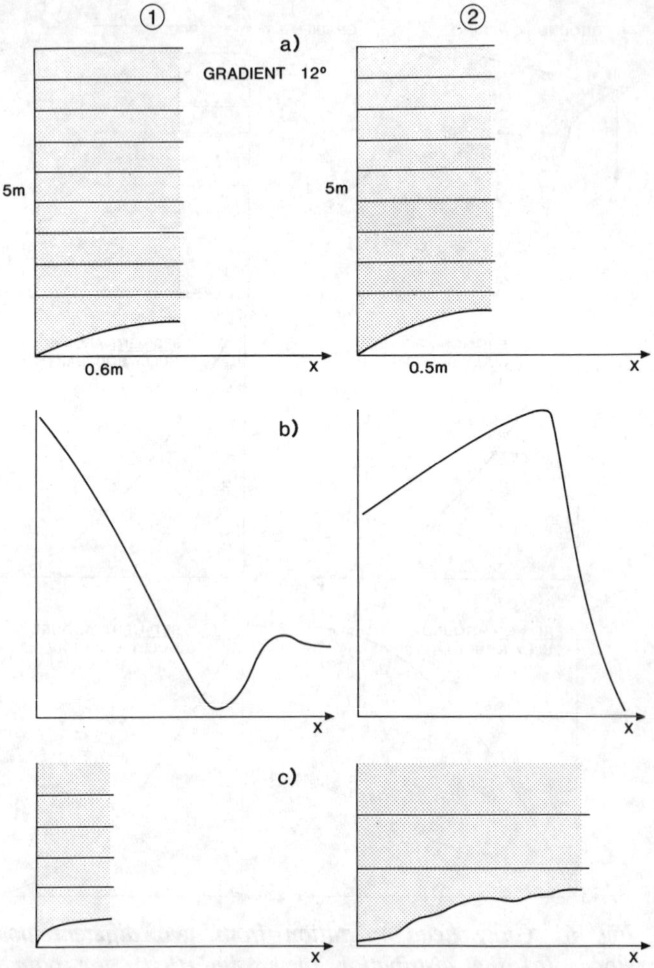

***Fig. 7** Two different gully heads migrating into identical planar surfaces: (a) plan view of slopes 1 and 2; (b) amount of erosion around gully heads 1 and 2; and (c) resulting migration of gully heads 1 and 2.*

flow has a greater erosive power at a distance from the apex of the gully head, but at the same time the same width of flow from upslope has to cross a greater width of gully boundary. There is a point near to the apex where the concentration of the flow leads to a local maxima of erosion. However, the trend is not strong and the gully head diminishes with time.

In the second case, however, the curvature of the gully head is greater, and the angle of the flowlines across the gully boundary changes more rapidly away from the apex. This has the effect of diminishing the erosive impact of the flow with distance from the central axis. At the same time, the erosive power of the flow is still increasing with distance downslope, and as a result, the focus of erosive power is strongly deflected away from the gully apex. A distinct bifurcation is observed at the end of the storm.

CONCLUSIONS

The implications of these simulations are that the balance between the erosional power in the flow and the relative concentration of this across the gully boundary may have a considerable influence on the way in which the gully will develop and branch. However, it is clear that the dynamics of sediment transport and the influence of slope profile may offset the effect of contributing area.

By modelling the dynamics of overland flow, as against using surrogates such as slope length or contributing area, certain characteristics of the erosive process can be explored. In shallow flows, with constant detachment and deposition, and where infiltration losses lead to a complex hydrology, there is no reason to expect that erosion rates are dependent on the distance from divide.

The investigation continues to focus on the influence of gully head geometry as the head migrates into planar, concave, and convex, converging and diverging slope forms. Two areas in particular are proposed for further development of the model:

(a) subsurface processes, in the context of both sapping and piping (e.g. Dunne, 1980); and

(b) gully head retreat as a function of slope instability along the gully boundary. The scale of these may vary from soil creep to slab or arcuate failures.

In the form presented here, this model provides a clear illustration of how the growth and bifurcation of a gully may be critically controlled by the way in which the gully head geometry and slope topography influence particular processes. It is proposed that this concept provides a useful framework for conducting digital and hardware simulations.

Acknowledgements I would like to thank Professor John Thornes for his inspiration and supervision, and Dr Ian Stewart, Computer Analyst, for his advice on the numerical algorithms. The financial support of the Natural Environment Research Council is acknowleged. I would also like to acknowledge the help and support of my former colleagues at the Department of Geography, Bristol University.

REFERENCES

Cordova, J. R., Rodriguez-Iturbe, I. & Vaca, P. (1983) On the development of drainage networks. In: *Recent Developments in the Explanation and Prediction of Erosion and Sediment Yield* (Proc. Exeter Symp. July 1986), 239-249. IAHS Publ no.137.

Dunne, T. (1980) Formation and controls of channel networks. *Progr. Phys. Geogr.* **4** (2), 211-239.

Dunne, T. & Aubrey, B. F. (1986) Evaluation of Horton's theory of sheetwash and rill erosion on the basis of field experiments. In: *Hillslope Processes.* Binghampton Symposia in Geomorphology no. **16** (ed A. D. Abrahams) 31-54.

Ellison, W. D. (1946) Soil erosion studies - Part One. *Agric. Eng* **28** (4), 145-146.

Foster, G. R. & Meyer, L. D. (1975) Mathematical simulation of upland erosion mechanics. *USDA Publication ARS-S-4*, 190-207.

Francis, C. F. (1985) The role and significance of surface and sub-surface hydrology on gully head growth in South East Spain. Unpublished PhD thesis, Univ. London, UK.

Horton, R. E. (1945) Erosional development of streams and their drainage basins. *Geol. Soc. Am. Bull.* **56**, 275-370.

Kibler, D. F. & Woolhiser, D. A. (1970) The kinematic cascade as a hydrological model. *Colorado State University Hydrology Paper* **39**, 27pp.

Kirkby, M. J. (1980a) Modelling water erosion processes. In: *Soil Erosion* (ed. M. J. Kirkby & R. P. C. Morgan), 183-216. Wiley, Chichester, UK.

Kirkby, M. J. (1980b) The stream head as a significant geomorphic threshold. In: *Thresholds in Geomorphology* (ed. D. R. Coates & J. D. Vitek) 53-74. George Allen & Unwin, London.

Moss, A. J., Green, P. & Hutka, J. (1982) Small channels: their experimental formation, nature and significance. *Earth Surf. Proc. Landforms* **7**, 401-415.

Musgrave, G. W. (1947) The quantitative evaluation of factors in water erosion: a first approximation. *J. Soil Wat. Conserv.* **2**, 133-138.

Schumm, S. A. (1956) *Evolution of Drainage Systems and Slopes in Badlands in Perth Amboy, New Jersey.* Office of Naval Research, Project NR389-042, Tech. Rep. 8. Columbia University.

Smith, T. R. & Bretherton, F. P. (1972) Stability and the conservation of mass in drainage basin evolution. *Wat. Resour. Res.* **8**, 506-524.

Thornes, J. B. (1976) Semi-arid erosional systems: case studies from Spain. *London School of Economics, Geographical Papers* no. 7.

Thornes, J. B. (1984) Gully growth and bifurcation. In: *Erosional Control - Man and Nature.* Proceedings of XV conference of the International Erosion Control Association. 131-140. Denver, USA.

Thornes, J. B. & Gilman, A. (1983) Potential and actual erosion around archaeological sites in South East Spain. In: *Rainfall Simulation, Runoff and Soil Erosion* (ed. J. De Ploey). *Catena Supplement* **4**, 91-113.

Zingg, A. W. (1940) Degree and length of land slope as it affects soil loss in runoff. *Agric. Engng* **21**, 59-64.

Erosion, Transport and Deposition Processes (Proceedings of the Jerusalem Workshop, March-April 1987). IAHS Publ. no. 189, 1990.

The dynamics of gully head recession in a savanna environment

EMMANUEL AJAYI OLOFIN
Department of Geography, Bayero University, Kano, Nigeria

Abstract Field measurements have been carried out to investigate the processes and rates of gully head recession in a tropical wet-and-dry location in Nigeria, over three wet seasons. A combination of gully wall erosion processes and gully floor basal sapping is found to enhance headscarp recession and gully widening through a cycle of undercutting, collapse and removal of collapsed material. The average recession rate was 2.32 m year^{-1}, and the fastest recession was 4.7 m year^{-1}. These rates amount to 8.6 m^3 year^{-1} and 20.4 m^3 year^{-1}, respectively, yielding an average loss of some 41 m^3 ha^{-1}. The major on-site factors accounting for these results include the amount of runoff arriving at each gully head, the texture and structure of the aeolian material into which gully heads are cut, and the absence of land management of the gullied zone. To stabilize the gullies, the recession cycle must be broken.

INTRODUCTION

Soil erosion, particularly in the Tropics, has been a vexed issue. Studies of soil erosion have highlighted the processes and factors that influence its occurrence, and its environmental effects, not the least of which is the loss of farmland (Olofin, 1978). Sheet wash and gullying comprise the two major types of erosion in humid and sub-humid areas.

In Nigeria, studies of soil erosion have been undertaken in both the humid area (Jeje, 1972, 1985; Ofomata, 1965, 1981) and the savanna zone (Ologe, 1972, 1973; Olofin, 1978, 1984). Rainfall erosivity is generally high in the Tropics. Courtney & Trudgill (1984) believe that up to 60% of rainfall is erosive, and Olofin (1980) contends that erodibility is high in soils composed mainly of silt and fine sand, while a high content of silt and clay results in large losses through mass movement. These hydrological and soil factors, together with slope angle, crop management and land conservation practices are the most important in a savanna zone such as that of the study area.

This paper presents the results of investigations into site factors and processes of gully head recession for a sub-humid savanna area of Nigeria. The data base spans three consecutive wet seasons, 1974 to 1976, with observations updated in 1986.

CLIMATE AND SITE CONDITIONS IN THE STUDY AREA

Figure 1 illustrates the location of the study area in the River Chalwa basin, Kano State, Nigeria. The area is characterized by a sub-humid savanna climate. The wet season lasts for about four months from June to September, sometimes beginning in the latter half of May. The mean annual rainfall is 860 mm, of which more than 35% may occur in August alone. The two wettest months, July and August, account for more than 60% of the rainfall.

The most relevant rainfall characteristic for soil erosion is rainfall intensity, which averages 30–40 mm hr^{-1} (Kowal & Kassam, 1978). Intensities of up to 80 mm hr^{-1} occur during rainstorms at the onset (May/June) and

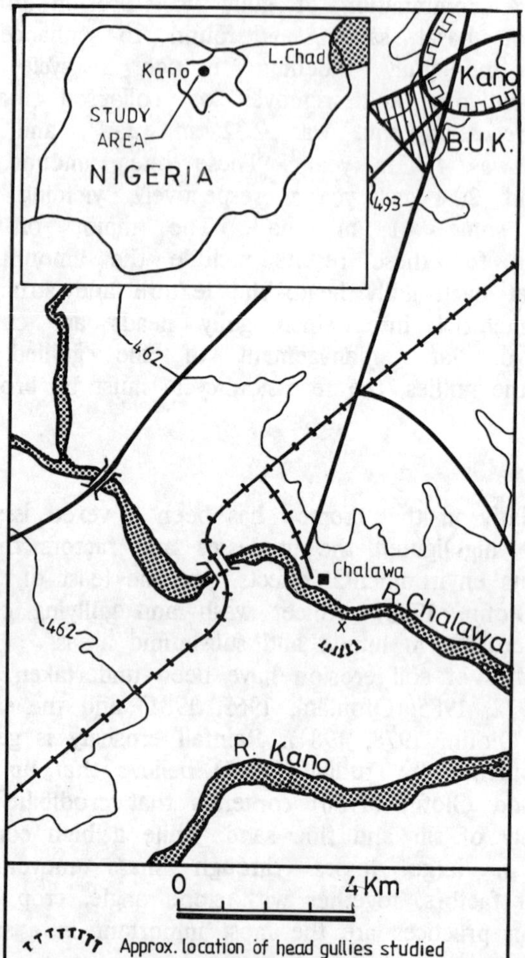

Fig. 1 Location of study site.

end (September) of the wet season (Leow & Olge, 1981), with storms lasting from 20–30 min. Storms occurring during the onset of the wet season, usually accompanied by winds in the range 50–70 km hr^{-1}, cause the most damage because the soil cover at that time is usually scanty, and cultivated fields have been ploughed.

The annual mean temperature is 27°C with an annual range of about 10°C. The warmest period occurs shortly before the rains in April and May when monthly mean temperatures of 30°C or more, and daily ranges less than 6°C are common. The coolest months are December and January when the tropical continental air mass, propelled by northeast Trade winds, prevails over the area. During this period the monthly mean temperature is about 22°C, with daily minima often dropping below 15°C, but maxima of over 30°C are quite common.

The site is underlain by crystalline rocks of Precambrian age which constitute the Basement Complex. Outcrops of these rocks, mainly igneous and metamorphic, occur everywhere on the landscape, including the bed of the Chalawa Channel. The channel itself is alluvial, 2 m deep and incised into an older broader channel about 240 m wide.

A terrace and upland plain, at the edge of which the gully heads are located, are overlain by aeolian material. The drift is about 1 m deep on the terrace where it covers past alluvial infills, except in the gully floors where it is completely absent. On the upland plains it is up to 2.5 m thick and overlies an ancient pediplain developed on the regolith of the Basement Complex rocks, the upper part of which has been covered by drift material (Fig. 2). Thus, the upland plains and terraces are depositional landforms whose mean slope angle ranges from 0–1.5°, separated by almost vertical scarps.

In the Nigerian Savanna aeolian material is believed to have "a modal

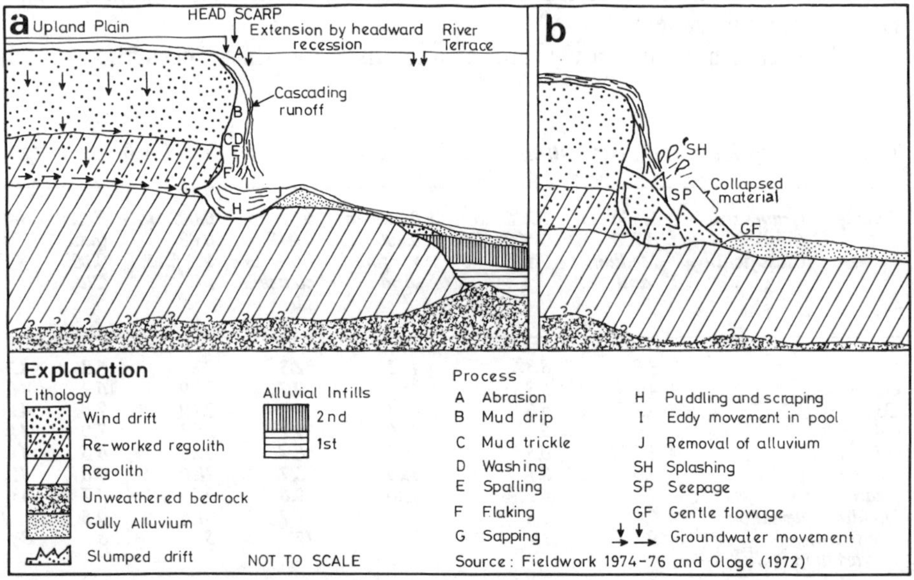

Fig. 2 Idealized stratigraphy of the site, and fluvial processes on (a) active and (b) stabilizing headscarps.

grain size in the fine sand to coarse silt range with a considerable percentage of clay size minerals" (Smith, 1982, p.39). The silt and clay fraction is estimated to be up to 50% by weight, clay alone accounting for 20–30% (Smith, 1982), although Ologe (1972) has pointed out that the material is coarser in the northern than in the southern parts of the area, and that both its thickness and extent decrease southwards. The study area is located in the northern part of the drift zone. The texture of the material at the site is illustrated in Table 1, which shows that the dominant grain size is silt to fine sand. The material easily crumbles once it is saturated but develops a crust under high intensity local rains. The grain size range also implies a high index of particle detachability under such intensities, particularly since the fines in this zone are believed to be removed from the top 10 cm of the soil profile (McTainsh, 1986).

For moderate rainfall intensities, sufficient water passes through the material, removing fine particles further down the profile and encouraging spring flow at the textural discontinuity existing between the aeolian material or reworked regolith layer and the regolith proper (Fig. 2). The discontinuity at the aeolian sediment/regolith interface occurs because the regolith is more compacted and has a higher clay fraction percentage.

Since 1979 dams have been constructed on the major tributaries of the Chalawa River upstream of the study site. These dams, and sand quarrying on the bed of the Chalawa, have initiated an incision into the channel of the river to a depth of 1.5 m and have resulted in a lowering of the stage by at least same margin. In response, gully floor and gully heads at the edge of the upland plain have been reactivated.

Land use at the site is characterized by rain fed cultivation of grains on the upland and terrace plains every year. Dry season market gardening cultivation is practised at some points at the edge of the low terrace, using a *shaduf* irrigation system.

The gullies under study are characteristic of those occurring on the

***Table 1** Particle size distribution of gully head aeolian material (% by weight)*

Sample	*Grain size (mm) Classification*	*0.500 to 0.750 Medium sand*	*0.200 to 0.500 Fine Sand*	*0.045 to 0.200 V. fine sand*	*0.02 to 0.045 Fine to coarse silt*	*0.002 to 0.02 V. fine silt*	*Finer than 0.002 Clay*	*Silt Clay %*
G1		*3.9*	*8.3*	*44.1*	*12.3*	*25.1*	*6.3*	*43.7*
G2		*4.1*	*5.3*	*45.7*	*9.7*	*28.9*	*6.2*	*44.8*
G3		*4.3*	*5.5*	*46.4*	*10.2*	*25.9*	*7.8*	*43.0*
G4		*3.6*	*5.6*	*46.0*	*9.7*	*28.1*	*7.0*	*44.8*
G5		*3.7*	*6.5*	*45.5*	*8.7*	*29.0*	*6.6*	*44.3*
G6		*4.6*	*8.1*	*46.0*	*6.7*	*28.0*	*6.6*	*41.3*
Mean		*4.0*	*6.6*	*45.6*	*9.6*	*27.5*	*6.7*	*43.8*
Standard deviation		*0.4*	*1.2*	*0.7*	*1.7*	*1.5*	*0.5*	*1.0*
Coefficient of variation (CV)		*9*	*18*	*2*	*17*	*5*	*8*	*3*

NOTE: The low CV values show that the aeolian material is approximately the same over the area.

valley sides of incised stream channels in the savanna areas of Nigeria. Such gullies are usually initiated on the vertical bends of the channels and work their way upslope by a process of headscarp recession (Ologe, 1972, 1973). As the channels themselves have been affected by polycyclic processes of incision and infilling under alternating pluvial and arid phases in the past (Smith, 1982), the current active gully heads are located at the boundary of the extended river terrace and the upland plains and are characterized by a steep and dissected upland edge.

The gully heads in this study are located 500 m from the incised channel of the Chalawa River, (Fig. 1), and are contained in a strip 500 m long and about 25 m wide at the edge of the upland plain.

METHODS OF STUDY

The unit studied was randomly selected from other units along the Chalawa River channel. Six active gully heads were considered. Field methods include the annual measurement of gully head dimensions in relation to fixed points, and the survey of gully plans at the end of each wet season (following the initial pre-wet season measurement in April 1974).

At each gully head, the following geometric variables were determined. Total top width (TTW) is the average (whole) top width of the gully in the receded section, such as line "AD" in Fig. 3, taking into account the pre-existing width.

Receded top width (RTW) is the average top width of each gully in the receded section such as lines "AB" and "CD" in Fig. 3, ignoring the pre-existing (BC) width. Base width (BSW) is the average bed width of each gully in the receded section, while the total depth (TDP) is the average depth in

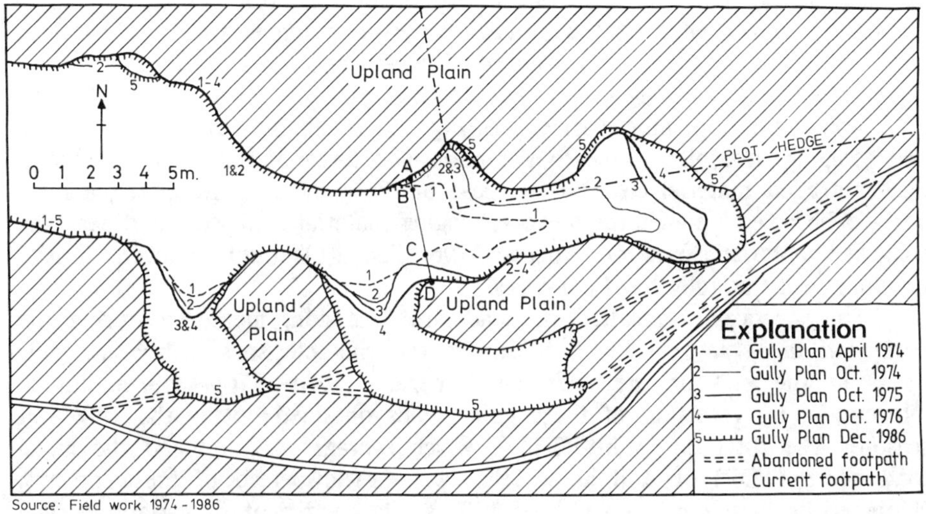

Fig. 3 Changing plan of head of Gully 1 between April 1979 and December 1986.

the receded section. The length of recession (RTL) is taken as the top length because headscarps are almost vertical, except for slight undercutting at the base. The changes in the plans and dimensions of the gully heads have been used to estimate the amount of sediment volume eroded (VER) and the surface area affected.

Field observations were also undertaken during the wet seasons to assess qualitatively processes and factors of headscarp recession. To this end, the land use, vegetation cover, effects of footpaths on runoff and actual erosion processes were observed.

All measurements and observations were made from April 1974 to October 1976. The escalation of sand quarrying and the construction of dams from 1976–1977 forced the cessation of observations since these activities introduced external factors. However, observations and measurements were updated in December 1986 when the most active gully head (Gully 1, Fig. 3) was surveyed and mapped. Samples of the sediment were also taken for all gully heads from a depth of 50–100 cm for particle size analysis.

HEADSCARP RECESSION PROCESSES AND GULLY HEAD CHARACTERISTICS

Field observations have revealed that the density, alignment and length of footpaths that lead into a gully head control the catchment area that drains into each gully head. Thus, different volumes of runoff are received by gully heads, even though both the upland plains slope angles and the rainfall intensity and amount are uniform over the sites. Although the runoff variable has not been quantified, it was observed that Gully 1 received the greatest amount of runoff during the period of study.

When runoff arrives at a receding gully head, the processes of soil erosion illustrated in Fig. 2(a) are set in motion. The runoff uses its load to abrade the upper edge of the headscarp before plunging to the gully floor where it creates an eddy pool. The eddy motion encourages flaking and basal sapping, aided by merging groundwater at the drift-regolith interface. On the gully wall, mud drip, mud trickle, washing and spalling occur, while puddling and scraping take place beneath the pool. The force of the plunging torrent pushes eroded material clear of the base of the gully wall, presenting a fresh base to be acted upon, until the overhanging material collapses or slides. The collapsed material slows the process down (Fig. 2(b)) until removed. If it is not removed, the gully head stabilizes.

These processes resulted in headscarp recession, measured during the period of the study. Table 2 illustrates the characteristics of the six gully heads at the end of the main study period (1974–1976) with additional information on recession of these gully heads obtained during the 1986 wet season. The total top width (TTW), which averages 4.85 m, and the mean total depth (TDP) of 2.18 m are the two least dispersed of the geometric characteristics; each has a value of 12% as its coefficient of variation. The mean volume eroded (VER, 25.66 m^3) was the most variable at 75%, while the mean receded top width (RTW) and base width (BSW) were also widely

Table 2 *Mean values of some geometric variables for selected gully heads*

Gully	*Base width (m)*	*Total top width (m)*	*Total depth (m)*	*Receded top width (m)*	*Receded top length (m)*	*Volume eroded (m^3)* *	*1986 Field check receded top length (one wet season) (m)*
1	*0.48*	*4.00*	*2.25*	*3.36*	*14.20*	*61.34*	*1.00*
2	*1.15*	*5.27*	*2.00*	*0.72*	*3.20*	*5.98*	*1.00*
3	*0.53*	*4.25*	*2.50*	*3.17*	*7.70*	*35.61*	*1.50*
4	*0.50*	*4.60*	*2.52*	*2.40*	*8.25*	*30.15*	*2.10*
5	*1.00*	*5.25*	*2.00*	*1.30*	*5.35*	*12.31*	*0.00*
6	*2.20*	*5.75*	*1.80*	*0.92*	*3.05*	*18.56*	*0.50*
Mean	*0.98*	*4.85*	*2.18*	*1.98*	*6.96*	*25.66*	*1.02*
Standard deviation	*0.61*	*0.62*	*0.27*	*1.22*	*3.80*	*19.37*	*0.67*
Coefficient of variation	*62*	*12*	*12*	*62*	*55*	*75*	*66*

* *Total volume eroded from all the gullies in three wet seasons = 154 m^3. The surface area actually affected = 190 m^2.*

variable (55, 62 & 52%, respectively).

Table 2 shows that the mean rate of recession during the study period was 2.3 m $year^{-1}$, which is considered rapid. The rate at which Gully 1 receded (4.7 m $year^{-1}$) is high. These rates are equivalent to the removal of an average of 8.6 m^3 $year^{-1}$ of sediment from each gully and 20.4 m^3 $year^{-1}$ from Gully 1 alone. Some 154 m^3 of earth was removed during the three wet seasons. The strip affected is about 1.25 ha. Thus, in the active gully zone 41 m^3 ha^{-1} $year^{-1}$ of sediment is removed by the gullying process alone.

Field surveys in 1986 showed that the rate of recession for that wet season was relatively slower than that of the study period in all the gully heads, except for Gully 4 which receded about 2 m (Table 2). However, the difference between the 1976 and 1986 Gully 1 plan (Fig. 3) shows that there must have been a period of episodic recession during the intervening decade in the region.

A statistical analysis relating some gully head characteristics to the rate of recession, indexed by the volume of material eroded (VER), provides the inter-relationships contained in Table 3.

Two variables stand out in their pattern of relationships. One is that TTW is negatively related to TDP, RTW, RTL and VER; but positively with BSW; all significant at, at least, 5% level. The second is the TDP which (in addition to its relationship with TTW) is negatively and significantly related to BSW, but positively (though insignificantly) related to RTW, RTL and VER. These r-values range between +0.58 and +0.67, indicating that with a larger sample size, these relationships could be valid. Other relationships can be read off Table 3.

DISCUSSION

In general, it is argued that valley side gullying in a savanna environment is

***Table 3** Pair-wise correlation matrix of gully head variables*

	BSW	*TTW*	*TDP*	*RTW*	*RTL*	*VER*
Base width (BSW)	*1*					
Total top width (TTW)	*+0.84**	*1*				
Total depth (TDP)	*-0.90**	*-0.83**	*1*			
Receded top width (RTW)	*-0.67*	*-0.83**	*+0.67*	*1*		
Receded top length (RTL)	*-0.73@*	*-0.89**	*+0.58*	*+0.75@*	*1*	
Volume eroded (VER)	*-0.70*	*-0.92***	*+0.62*	*+0.81**	*+0.98****	*1*

Degrees of freedom (n - 2) = 4.
**** Significant at 0.1% level,*
*** Significant at 1.0% level.*
** Significant at 5.0% level.*
@ Possible valid relationship (significant at 10%).

initiated by the incision of the main river channel (Ologe 1972, 1973), and that the depth of gully incision is set by the local base level (Smith, 1982) provided by the height of flow in the main channel (Olofin, 1984). The recent channel incision, resulting in the reactivation of gully floors, underlines the importance of the control exerted by the main channels on gully side erosion. Thus, the gullies were probably initiated during the first incision of the Chalawa channel; headscarps working their way upslope to their current location.

However, the present study has shown that other environmental factors and human interference controlled the growth of the gullies once they had been initiated. Prominent among them are the lithology of the surface material; the pattern and intensity of rainfall, the amount of runoff reaching the gully heads, land use, and land management. While each factor may be isolated for discussion, it should be noted that they can be additive in terms of effect.

The texture of the sediment referred to above is dominated by silt to fine sand particles, which develop crusts under high rainfall intensity, but which still permit sufficient water to infiltrate during rainfalls of moderate to low intensities to initiate spring flow at the drift-regolith interface. This flow results in basal sapping of headscarps similar to that observed by Piest *et al.* (1975) for vertical loess banks and head cuts in Iowa.

Since the texture and depth of the drift are relatively uniform at all the gully heads, a uniform effect should be expected, more so since rainfall inputs and slope conditions are also uniform. Uniformity in these factors appears to control the similarities in total gully depth and width at the "mature stage", but not the characteristics of the active headscarp (such as the rate of incision) which are very variable. This author contends that site factors influence such variation.

One such factor is the amount of runoff that reaches the headscarps. As mentioned earlier, this variable was not quantified. However, a qualitative evaluation suggests that Gully 1 (Fig. 3) receives at least double the amount of runoff reaching the other gully heads because it has the largest drainage area, thus explaining its very rapid rate of recession.

Coupled with the amount of runoff are land use and management practices. Land use in the study site lays the land bare at the onset of rains, when the greatest erosion occurs. In many cases, cultivation is carried right to the edge of gully heads, and footpaths which represent the divides within the flat upland terrain, run up to the edge of the gully heads. The effect of the land use/management factors differ from one gully head to the other. Variation in the rate of recession, volume eroded and other gully head characteristics relate directly to their combination. In other words, gully heads that receded the most (Gully 1, for example) are those where the combination of land use and management factors present the greatest risk of soil erosion.

It also appears that the characteristics of pre-existing gullies near the receding headscarp provide a set of dynamic or intrinsic factors that influence the rate of recession and the volume of material removed. These factors can explain, in part, the relationships described earlier (Table 3). The wider gully heads recede mainly laterally, with limited headward cuts, while the shallower ones recede without basal undercutting (which would have accelerated the process) because their floors are still within the sediment layer. However, the wider gully heads are also the relatively shallower ones. It follows then that in addition to the factors already described, differential rates of recession and related characteristics are also explained by a dynamic factor associated with the differences in the geometric characteristics of the older sections of the gully heads, in so far as they affect the cycle of undercutting, collapse and removal of gully head material.

CONCLUSION

Although the incision of the main channel initiates valley side gullying in the area, the rate of growth of incipient gullies is dependent on other environmental and anthropogenic factors. Much of the eroded material is lost at the onset of the wet season each year, mainly through mass wasting arising from a sequence of undermining, collapse of undermined materials and the subsequent removal of collapsed material. It is estimated that some 41 m^3 ha^{-1} $year^{-1}$ is lost in this way.

This soil erosion rate constitutes a great loss of agricultural land and should be brought under control. It appears that in order to inhibit or completely stop the process of gully head recession in the study area (and in similar environments), the sequence must be broken. The most appropriate point to intervene is the removal stage. Once the removal of the collapsed material is prevented (through sediment traps and other methods), undermining, flaking and washing will be checked and subsequent collapse can be prevented. In addition, runoff reaching the gullies should be reduced in velocity and in quantity through the creation of grass barrier strips and other obstacles around the gullies, and by preventing footpaths from running into gully heads. The resultant increase in infiltration will not increase basal sapping if removal of collapsed material has been prevented. If this is not done, increased infiltration will do more harm than good in this type of situation. Also, the gully walls and headscarps should be treated to produce gentler slopes, especially in urban settings, in the manner described by Olofin (1987). Man must also develop better land management practices in the area.

It should be noted that results presented in this work and the techniques suggested relate only to aeolian-covered plains of the sub-humid savanna zone of Nigeria, and areas with similar hydrological and site characteristics. Even in such areas more extensive and quantified research is required before firm conclusions can be made about the dynamics of gully head recession.

Acknowledgements I thank Mr John F. Antwi of the Department of Geography, Bayero University, Kano, for drawing the diagrams, and Mr Olayinka F. Agbeniyi of the Department of Hydrology and Water Resources Management, University of Agriculture, Abeokuta for typing the final manuscript.

REFERENCES

Courtney, F. M. & Trudgill, S. T. (1984) *The Soil: An Introduction to Soil Study.* E. Arnold, London.

Jeje, L. K. (1972) Landform development at the boundary of sedimentary and crystalline rocks in southwestern Nigeria. *Trop. Geogr.* **34**, 25-33.

Jeje, L. K. (1985) Runoff and soil loss from erosion plots in Ife area of south-western Nigeria. In: *Abstracts of papers for First International Conference on Geomorphology* (ed. T. Spencer), 297. Univ. Manchester, Manchester, UK

Kowal, J. M. & Kassam, A. H. (1978) *Agricultural Ecology of the Savanna.* Oxford Univ. Press, Oxford, UK.

Leow, K. S. & Ologe, K. O. (1981) Rates of soil wash under a savanna climate, Zaria, northern Nigeria. *Summaries of Papers, 24th Conf. Nigerian Geographical Association (NGA)* 18-22.

McTainsh, G. H. (1986) A dust monitoring programme for desertification control in West Africa. *Environ. Conserv.* **13**, (1) 17-25.

Ofomata, G. E. K. (1965) Factors of soil erosion in the Enugu region of Nigeria. *Geogr. J.* **8**, 45-59.

Ofomata, G. E. K. (1981) The management of soil erosion problems in Nigeria. Invited paper presented at the 24th Conference of the NGA, Kano.

Olofin, E. A. (1978) Effects of gully processes on farmlands in the savanna areas of Nigeria:Chalawa basin case study. *Kano Studies* NS1 (3), 74-83.

Olofin, E. A. (1980) The determination and significance of indices of soil erodibility, Ulu Langat District, Selangor: a case study. *Malaysian J. Tropical Geogr.* **2**, 26-34.

Olofin, E. A. (1984) Some effects of the Tiga Dam on valleyside erosion in downstream reaches of the River Kano. *Appl. Geogr.* **4**(4), 321-332.

Olofin, E. A. (1987) Restoration and protection of degraded terrain. In: *Ecological Disasters in Nigeria: Drought and Desertification* (eds. Sagua, V.O., Ojanuga, A.U., Enabor, E.E., Mortimore, M.J., Kio, P.R.O. & Kalu, A.C.) 323-334, Federal Ministry of Science and Technology, Lagos.

Ologe, K. O. (1972) Gullies in the Zaria area: a preliminary study of headscarp recession. *Savanna* **1**(1), 55-57.

Ologe, K. O. (1973) Some aspects of the problem of modern gully erosion in the northern states of Nigeria. Paper presented at the HTC symposium on Hydrology and Water Resources Development in Nigeria, Zaria.

Piest, R. F., Bradford, J. M., & Wyatt, G. (1975) Soil erosion and sediment transport from gullies. *J. Hydraul. Div ASCE,* **101**(1), 65-80.

Smith, B. G. (1982) Effects of climate and landuse changes on gully development; an example from northern Nigeria. *Z. Geomorphol. Suppl.* **44**, 33-51.

Erosion, Transport and Deposition Processes (Proceedings of the Jerusalem Workshop, March-April 1987). IAHS Publ. no. 189, 1990.

Conditions for the evacuation of rock fragments from cultivated upland areas during rainstorms

J. W. A. POESEN

Research Associate, National Fund for Scientific Research, Laboratory of Experimental Geomorphology, K. U. Leuven, Redingenstraat 16 bis, B-3000 Leuven, Belgium

Abstract Most studies dealing with erosion of stony soils have treated rock fragments at the soil surface as mulch elements. This study considers rock fragments on upland areas as erodible particles and addresses the following questions: (a) which erosion processes are capable of moving rock fragments? (b) under what hydraulic conditions do rock fragments start moving on slopes? (c) what factors determine displacement distances of rock fragments? The monitoring of coloured rock fragment movement revealed that during a moderate rainfall event, rock fragments up to 9.0 cm in diameter travelled downslope by rill flow. The competence of interrill flow was about one order of magnitude smaller. Incipient motion conditions for rock fragments lying on a rill bed coincide with a critical Shields entrainment parameter (θ_c) of 0.012 rather than with θ_c = 0.06. Rock fragment transport distance was controlled more by fragment size than by fragment shape and it correlated better with rill bed slope than with peak rill flow discharge.

INTRODUCTION

In literature dealing with soil erosion on cultivated lands the presence of rock fragments, i.e. particles 2 mm or larger in diameter and including all sizes that have horizontal dimensions less than the size of a pedon (Miller & Guthrie, 1984), on the soil surface or in the plough layer is usually considered to be beneficial insofar as the rock fragments reduce the intensity of soil degradation processes such as surface sealing, compaction, interrill as well as rill erosion. In fact rock fragments lying on the soil surface are regarded as mulch elements, reducing or even eliminating raindrop impact energy - which leads to a retardation of surface sealing and, hence, to an increase of water intake into the soil surface - and reducing runoff velocities. Due to the latter, the detaching and transporting capacity of overland flow will also be reduced.

These aspects are well documented by several studies (Seginer *et al.*, 1962; Adams, 1966; Epstein *et al.*, 1966; Deffontaines & De Montard, 1968; Meeuwig, 1970; Meyer *et al.*, 1972; Collinet & Valentin, 1984). Very few studies, however, exist in which rock fragments are considered as erodible particles. Field observations on agricultural lands in central Belgium reveal that considerable amounts of rock fragments can be eroded from cultivated upland areas and, in part, be deposited at the foot slopes (Fig. 1). A field study and a laboratory experiment were thus set up to answer the following questions:

(a) Which erosion processes are capable of moving rock fragments at the surface of stony soils?

(b) Under what conditions do rock fragments start moving on slopes?

(c) What factors determine displacement distances of rock fragments?

Fig. 1 Recent colluvial deposits containing considerable amounts of rock fragments (Huldenberg, central Belgium). Length of stick equals 60 cm.

MATERIALS AND METHODS

Field study

The field study was conducted on a 0.75 ha field plot, located in Huldenberg (central Belgium) and is described in detail in Govers & Poesen (1986). Soils on the plot can be described as gravelly/cobbly loam or gravelly/cobbly sandy loam. Rock fragments in the plough layer or at the soil surface originate from gravel-rich fluviatile deposits (mainly flint pebbles and cobbles), probably of early Quaternary age, at shallow depth. They are brought to the surface either by surface lowering due to soil erosion, by ploughing, or by freezing and thawing.

After the field plot was tilled and placed in a conventional seed bed on 15 November 1983, the soil surface was kept bare by the application of herbicides, replicating to some extent semi-arid conditions. As a consequence, surface sealing, compaction of the plough layer and inter-rill as well as rill erosion occurred the following year (Poesen & Govers, 1986; Fig. 2). Due to the selective removal of fines, an erosion pavement developed on the inter-rill areas while, at the bottom of the rills, a discontinuous layer of rock fragments was deposited after hydraulic erosion of the rill bed and/or mass movement processes on the rill banks took place. On 7 November 1984, four rills were selected on the basis of catchment size and the dimensions of the rill cross section and, in addition, five seeding transects were chosen on the basis of surface slope gradient (Fig. 3). Each seeding transect was named after the colour of the painted rock fragment used. Following this, 687 coloured and numbered flint pebbles and cobbles, with intermediate diameters ranging from 0.35 to 9.8 cm and having both variable roundness and shape (flatness index (FI) varied between 1.1 and 3.9; $FI = (L + I)/(2S)$ with L denoting the longest, I the intermediate and S the shortest dimension of the fragment along three perpendicular axes), were placed on the rill beds at selected sites. Each rill site was located close to each crossing of a seeding transect on one of the four selected rills. At each rill site, three places, about

Fig. 2 The 170 m long and 45 m wide rilled field plot in Huldenberg (central Belgium).

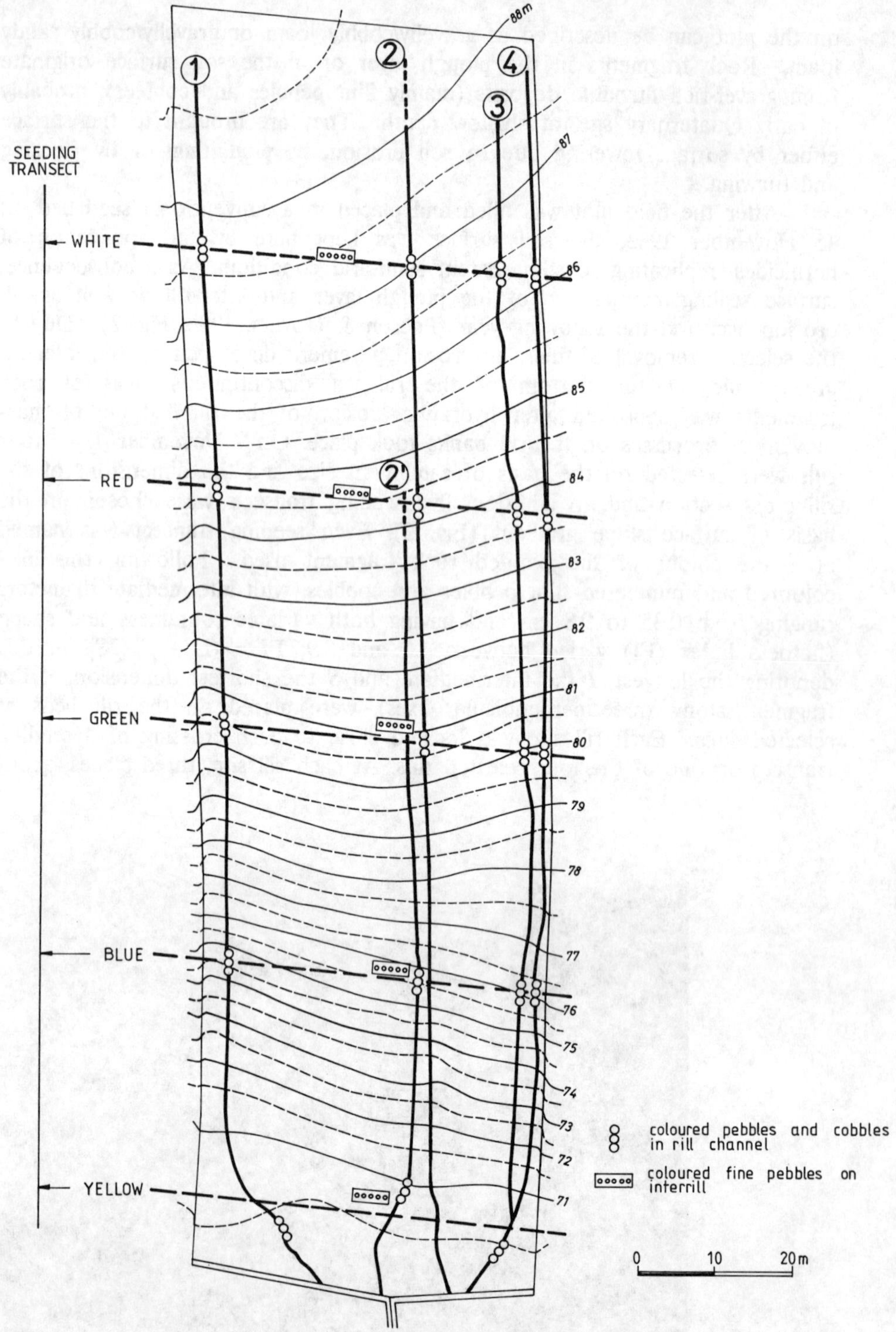

Fig. 3 Topographic map of the Huldenberg field plot and location of the selected inter-rill and rill sites.

1 m apart, were selected and a set of 5 to 35 coloured fragments was randomly placed in a cluster on the rill bed. Care was taken to ensure that rock fragments were oriented in a manner commonly adopted by undisturbed rock fragments of the same type.

The drainage area (A) of each rill upslope of a selected rill site was calculated after mapping flowlines, visible on the inter-rills, on a topographic map of scale 1:200. Area A varied between 2.4 and 159.2 m^2. Rill bed width ranged from 3 to 52 cm and the slope gradient of the rill beds varied between 0.049 and 0.268. In addition, coloured fine pebbles with intermediate diameters ranging from 0.2 to 0.8 cm, and with variable flatness, were placed on the inter-rill area along 1 m long seeding lines, close to the transects. A recording rain gauge, installed at a distance of 650 m from the field plot, provided the necessary rainfall data. On 26 November 1984, following a rainy period, the coloured rock fragments in the rills and on the inter-rills were recovered. Some 71% of the fragments used as tracers were found again.

Laboratory experiments

Threshold conditions for incipient motion of rock fragments by rill flow were determined in the laboratory using a tilting flume which consisted of a 12 m long PVC roof gutter with a trapezoidal cross section (Fig. 4). The bottom width of the flume was 11.8 cm, typical for rills in the field, while the walls were inclined at 0.176. The flume slope gradient could be varied from 0 to 0.50.

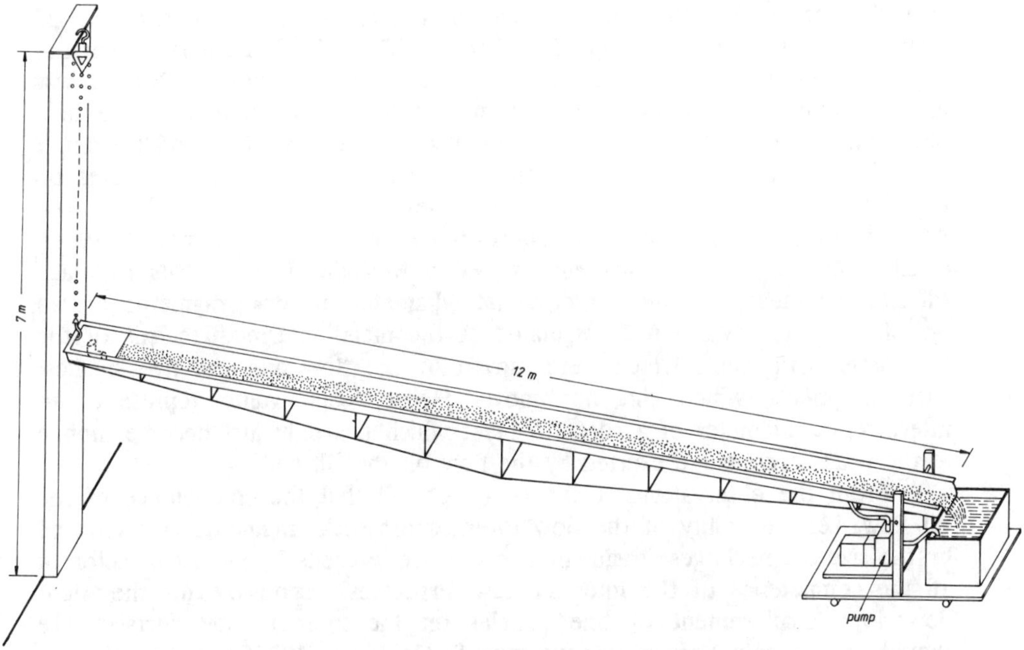

Fig. 4 Experimental setup in the laboratory.

To simulate different rill bed roughnesses, very fine sand (D_{50} = 0.010 cm, i.e. a smooth bed) and medium gravel (D_{50} = 0.6 cm, i.e. a rough bed) were fixed with a hard water-repellent glue to the bottom of two flumes. Rill flow was simulated by recirculating tap water with a temperature varying between 15 and 22°C. Maximum unit discharge (q) equalled 135 $cm^2 s^{-1}$. Rollwave-trains were eliminated by means of a net suspended in the flow at about 5.5 m upstream from the bottom end of the rill channel.

Twenty-four flint (ρ_s = 2.65 g cm^{-3}) pebbles and cobbles ranging in intermediate diameter from 0.3 cm to 6.6 cm and having different FI values were used (De Wilde, 1986). After establishing uniform flow close to the threshold condition for a given fragment size, the individual fragment was gently placed on the rill bed in its most stable position. For a given unit discharge, the slope was adjusted while making simultaneous observations for any movement of the fragment. Each fragment was tested for at least 10 points along the last 4 m of the rill channel. When the ratio between the number of unstable spots to the total number of tested spots fell between 0.75 and 0.85, these values being arbitrarily chosen, incipient motion was considered to be reached. At that moment flow discharge, mean flow velocity (using coloured dye), water temperature and rill bed gradient were measured.

RESULTS AND DISCUSSIONS

Inter-rill and rill flow competence

During the period of observation (7 November to 26 November 1984) total rainfall equalled 63 mm with a maximum rainfall intensity occurring on 22 November of 6.0 mm in 12 min (or 30.0 mm h^{-1}). Such a rainfall event has a return period of 6 months in central Belgium (Laurant, 1976). It was assumed that the main movement of the traced rock fragments on the field plot could be attributed to the peak flow occurring during the rainfall event.

During the period of observation, movement of the largest pebbles on the inter-rills, i.e. fragments with an intermediate diameter of 0.8 cm, occurred over a maximum downslope distance of 5 cm. In the rills, however, considerably larger rock fragments moved downslope. Table 1 lists for each rill site the mean maximum intermediate diameter of rock fragments moved by rill flow. This value was calculated as the mean intermediate axis of the five largest fragments which were moved by the flow and the five smallest rock fragments which did not move. Hence, each value represents an intermediate diameter of a rock fragment which has only just become mobile at the peak stress value exerted by the flow on the rill bed.

From the observations it can be concluded that the competence of the rill flow, i.e. the ability of the flow to transport rock fragments as measured by the size of the largest fragment it can move, exceeds by almost a factor of 10 the competence of the inter-rill flow. Processes responsible for the slight downslope displacement of fine pebbles on the inter-rill soil surfaces are thought to be splash-creep (Moeyersons & De Ploey, 1976) and runoff-creep (De Ploey & Moeyersons, 1975). Assuming a uniformly distributed sheet flow

***Table 1** Mean maximum intermediate diameter (cm) of rock fragments, moved by rill flow during the period of observation*

Transect	*Rill no.*			
	1	*2*	*3*	*4*
white	1.2	6.2	2.6	5.7
red	4.3	7.8	5.4	9.0
green	4.2	6.1	6.0	6.3
blue	5.7	5.8	7.5	5.8
yellow	4.6	5.4	7.0	

on the inter-rill areas of the field plot, one will find that during peak runoff sheet flow unit discharge never exceeded 10 cm^2s^{-1}. Such flow discharges are usually not competent to transport rock fragments with diameters ranging between 1 and 8 cm, as shown by the experimental results of De Ploey & Moeyersons (1975). On the other hand, rill flow caused by a moderate rainfall event is capable of moving rock fragments which have mean intermediate diameters of up to 9 cm (Table 1). Hence, rill flow and other forms of concentrated overland flow (e.g. ephemeral gully flow) can be held responsible as the most important processes evacuating pebbles and cobbles from upland areas.

Threshold conditions for incipient motion of rock fragments by rill flow

Data on the hydraulic threshold conditions for the incipient motion of sediment particles in turbulent flow are usually expressed in terms of Shields' model (Vanoni, 1977):

$$\theta = \frac{\tau_0}{(\rho_s - \rho)gd} = f(u_*d/\nu) = f(Re_*) \qquad (1)$$

in which:

θ = Shields entrainment parameter;
ρ_s, ρ = density of sediment, density of fluid (kg m^{-3});
g = acceleration due to gravity (m s^{-2});
d = effective diameter of a bed particle in a state of incipient motion (m);
τ_0 = boundary shear stress (N m^{-2}); (= ρgRS with R = hydraulic radius (m) and S = rill bed slope gradient);
f = function of;
u_* = shear velocity (= $(\tau_0/\rho)^{1/2}$);
ν = kinematic fluid viscosity (m^2 s^{-1}); and
Re_* = grain Reynolds number.

θ must be assigned a critical value (θ_c) in order to solve the left hand term of equation (1) for a given grain diameter (*d*) or stress value (τ_c). Hence, *d* is a measure of theoretical flow competence.

In order to compare our field data on rock fragment movement initiation by rill flow to existing theory, θ_c and Re_* were calculated for the different rill sites. Since the time of concentration for each rill catchment was well below the duration of the rainfall event causing peak runoff (i.e. 12 min), rill flow discharge was calculated by the rational formula:

$$Q = C\,I\,A \qquad (2)$$

with: Q = rill flow discharge ($m^3\ s^{-1}$);
C = runoff coefficient;
I = rainfall intensity ($m^3\ s^{-1}\ m^{-2}$); and
A = rill catchment area (m^2).

On compacted and sealed soils of the field plot, having a moisture content exceeding field capacity, Govers (1986) obtained *C* values approaching 1.00 when rainfall intensity equalled several tens of mm per hour. Hence, for our calculations, *C* was set equal to 1.0. Next, unit peak rill flow discharge was calculated as $Q(b_s)^{-1}$, where b_s represents the smallest rill bottom width for each rill site. The b_s value was chosen in order to obtain the maximum possible peak unit rill flow discharge. Calculated *q* values varied between 2.4 and 99.7 $cm^2\ s^{-1}$. Hydraulic radius (expressed in m) was then calculated using a modified Manning formula:

$$R = (q\ n\ S^{-0.5})^{0.6} \qquad (3)$$

with *q* expressed in $m^2\ s^{-1}$ and n = Manning roughness coefficient.

The value of *n* was estimated on the basis of photographs of the rill beds and published *n* values (Foster *et al.*, 1984) as well as measured *n* values for different rill beds in the laboratory (De Wilde, 1986). For the rill beds at the field plot, *n* varied between 0.015 for a flat sandy bed and 0.05 for an irregular gravelly bed.

Using ρ, ρ_s, q, R, S, ν and d (taken from Table 1) as input data, θ_c and Re_* were calculated using equation (1). Fig. 5 shows the field rill data plot well below $\theta_c = 0.06$ for rough turbulent flow conditions ($Re_* \geqslant 400$); i.e. on average $\theta_c = 0.012$. In addition, these data show a significant positive relation between θ_c and Re_* which can be well described by a power relation with exponent 0.41. A graphed representation of the laboratory data, using for *d* the rock fragment diameter parallel to the flow instead of the intermediate diameter, is shown in Fig. 6.

Laboratory data corresponding to the rough rill bed plot close to the Shields curve, but data corresponding to the smooth rill bed plot well below $\theta_c = 0.06$ and can be represented by an equation very similar to that found for the field data.

From the laboratory data it can be concluded that incipient motion conditions for rock fragments are determined, to a large extent, by the roughness of the rill bed. For a rough rill bed the data almost coincide with the Shields curve, while for a smooth rill bed incipient motion conditions are

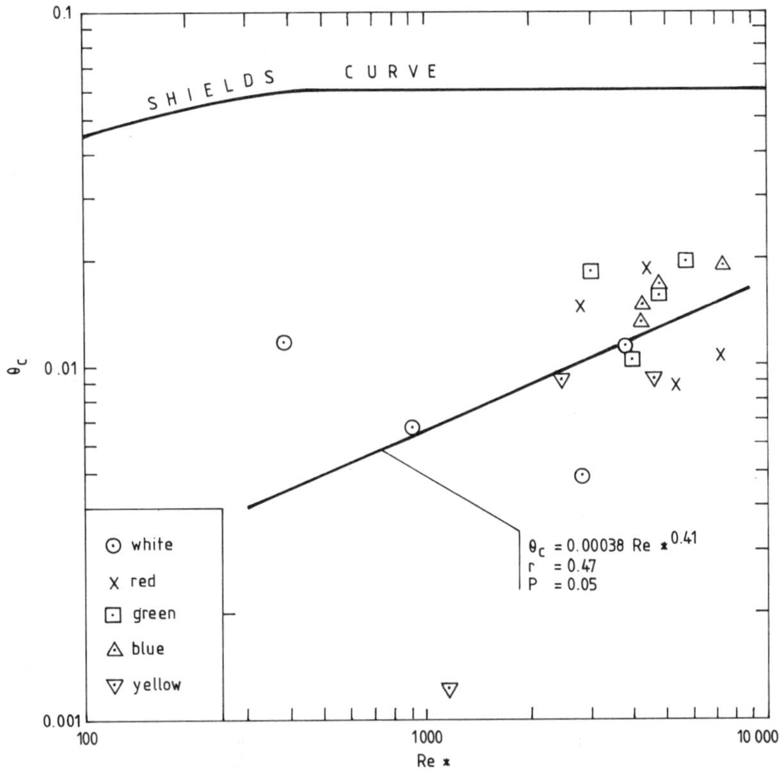

Fig. 5 Shields' entrainment parameter (θ_c) versus grain Reynolds number (Re_*) for the field rills. Different symbols refer to different seeding transects.

reached at θ_c values which, on average, equal one-fifth of the θ_c value given by Shields. In order to develop his model, Shields used results from almost equi-dimensional grains laid in flat beds. This was clearly not the case in our field measurements given that rock fragments were laid on a relatively smooth rill bed. Accordingly, these fragments protruded considerably above the mean bed elevation. Experiments by Fenton & Abbott (1977) and field data compiled by Andrews (1983) clearly show that the entrainment parameter decreases with an increasing degree of exposure of individual grains to the fluid flow.

On the basis of the laboratory findings, it can be stated that the field data are not well represented by a classical Shields entrainment parameter of 0.06 because the traced rock fragments were placed *on* a relatively *smooth* rill bed. Consequently, the angle of repose was much lower than for a situation in which the fragments would have been placed *in between* fragments of the same size. This analysis clearly shows that the threshold relation for the transport of rock fragments by overland flow on uplands is different from those developed for rivers. Similar findings were recently reported by Abrahams *et al.* (1988) for the transport of sediment by overland flow on desert hillslopes.

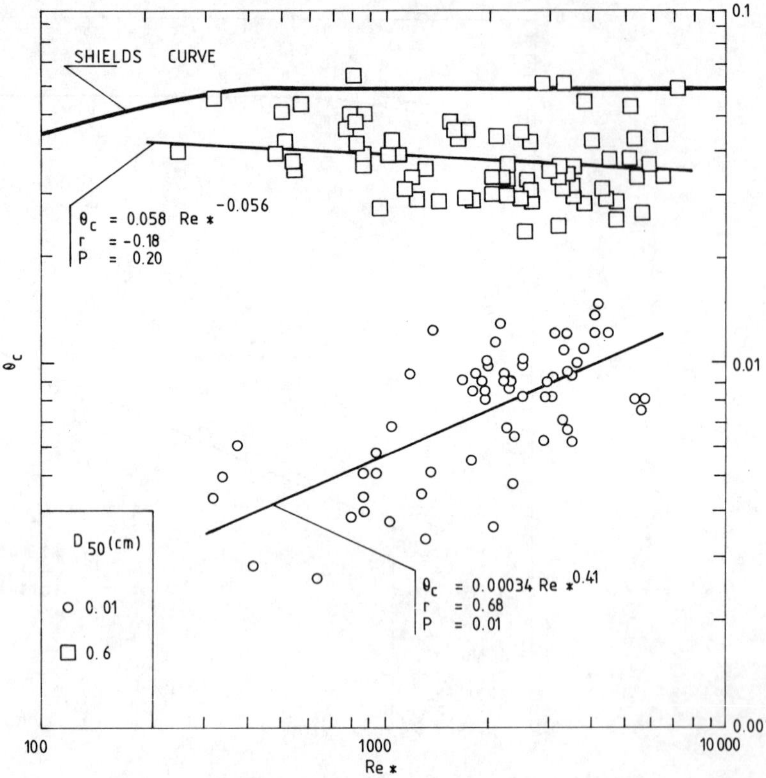

Fig. 6 Shields' entrainment parameter (θ_c) versus grain Reynolds number (Re_*) for the laboratory rills with a smooth (D_{50} = 0.01 cm) and a rough (D_{50} = 0.6 cm) bed.

Factors controlling displacement distances of rock fragments in rills

The factors investigated controlling displacement distances were divided into two groups; rock fragment properties and rill site characteristics.

Properties of rock fragments investigated were size (intermediate diameter) and shape (flatness index). Figure 7 illustrates the relationship between fragment size, fragment shape and transport distance for two selected rill sites. In general, fragment size and distance moved tend to be inversely related (e.g. Fig. 7, blue transect, rill 3). For a given fragment diameter, however, a large variation in transport distances exists. This variation decreases as size increases.

A visual inspection of all scatter diagrams leads to the conclusion that there is little dependency of travel distance on shape. Thus it can be concluded that fragment size plays a more important role with respect to the displacement of rock fragments than does fragment shape. However, there is no defined relationship between rock fragment size and transport distance. This is in accord with findings of Leopold *et al.* (1966) and Schick *et al.* (1987) for ephemeral streams. The relatively aselective nature of rill flow can partly be explained by the stochastic nature of gravel entrainment and by rill

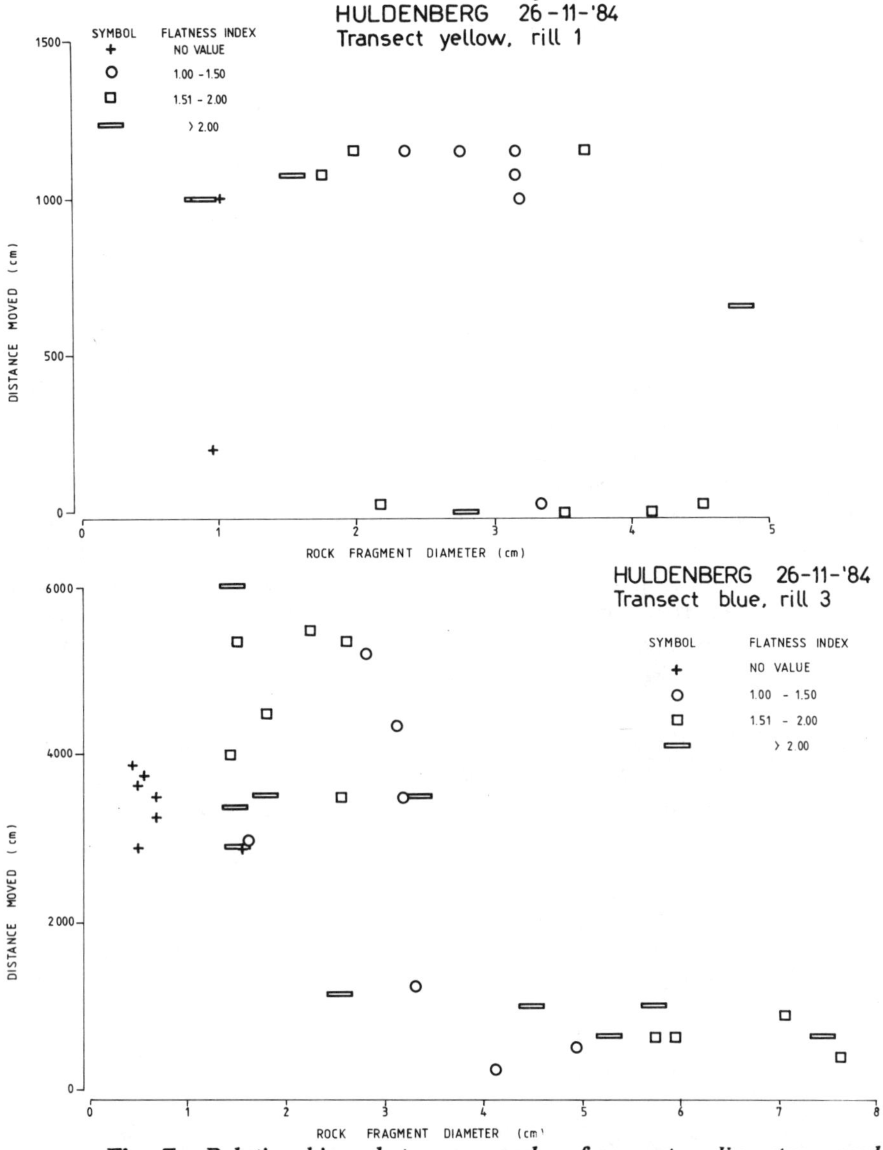

Fig. 7 Relationship between rock fragment diameter and displacement distance in rills for two selected sites.

bed roughness; pools formed in the rill beds often trapped considerable amounts of rock fragments. This is illustrated in Fig. 7 (yellow transect, rill 1) where a set of different sized particles (1 to 3.5 cm) were trapped in a rill bed depression occurring at a distance of 10 to 12 m downslope of the seeding transect.

In order to explain the variation in observed rock fragment transport distances between the different rill sites, we deduced for each rill site the mean displacement distances, corresponding to a 1 cm (Y1) and a 4 cm (Y2) diameter rock fragment from the scatter diagrams using linear regression equations. Hence, part of the variance in displacement distance due to rock fragment diameter was removed. Next, both displacement distances were related to peak rill flow unit discharge (*q* was calculated with the rational formula) and mean slope gradient of the rill bed (*S*). From the analysis it could be concluded that rock fragment displacement distances always correlate better with *S* ($r = 0.43$ to 0.44) than with *q* ($r = 0.27$ to 0.33). In addition, critical bed slope angle (S_{cr}), for rill flow transport of rock fragments with diameters between 1 and 4 cm, varied in our study between 2° 20' and 3° 30'.

These S_{cr} values were obtained by extrapolating the curve, fitting data points in a (*S*)–(*Y*1,*Y*2) diagram for each of the four selected rills (e.g., Fig. 8). This observation indicates that incision of rills in stony soils can commence during moderate rainfall events on hillslopes having gradients above these S_{cr} values. These critical slopes are in agreement with reported S_{cr} values for initiating rill and gully formation on fine colluvial gravels (i.e. 2°, Newson, 1980) as well as on loamy soils (2–3°, Savat & De Ploey, 1982). In addition, these observations indicate that complete surface armouring, due to selective erosion of fines and the concentration of rock fragments at the surface, will essentially occur on slopes less than the S_{cr} values mentioned. For values exceeding critical slopes, the probability of complete surface armouring decreases.

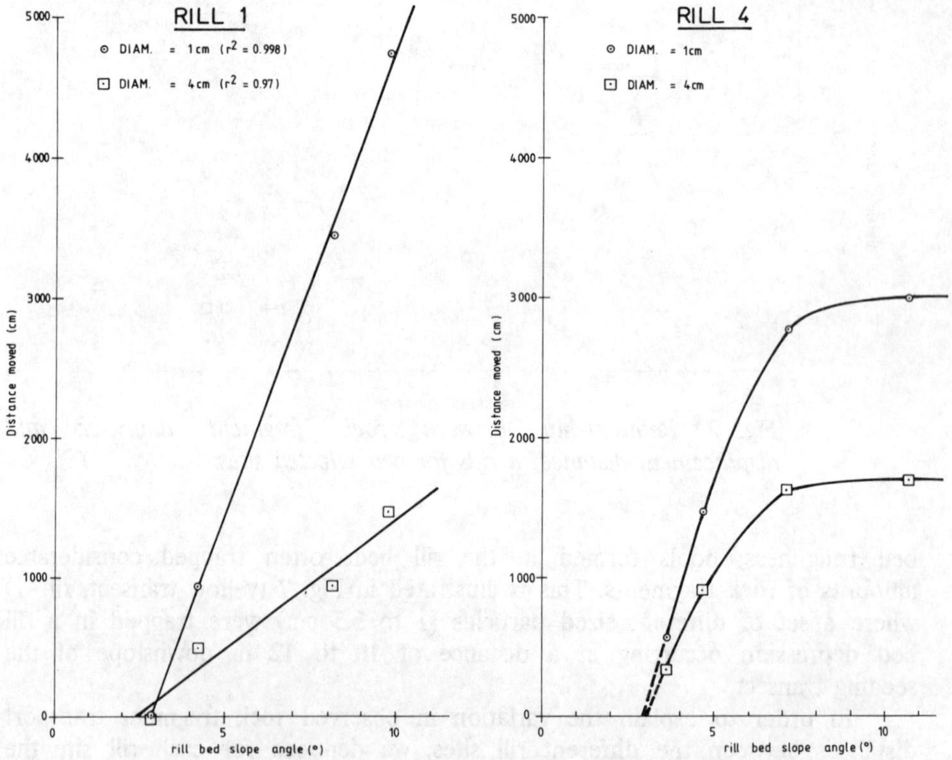

***Fig. 8** Relationships between slope angle of rill bed and mean transport distance for two selected pebble diameters.*

Implications of results

(a) Rock fragment content is a soil property. In addition, rock fragments themselves can be eroded by rill flow during moderate rainfall events, as shown in this study. Hence, it would be scientifically more accurate to include the effects of rock fragments on soil loss in a "universal soil loss equation soil erodibility factor" (K) rather than in a "cover and management factor" (C) (Wischmeier & Smith, 1978; Box & Meyer, 1984).

(b) This study clearly demonstrates that under moderate rainfall conditions rock fragments, having intermediate diameters up to 9 cm, can be transported downslope over considerable distances by rill flow. Hence, when applying stones as a mulch for erosion control on rillable soils (e.g. Adams, 1966; Meyer *et al.*, 1972; Jennings & Jarrett, 1985; Kochenderfer & Helvey, 1987), attention should be paid to selecting rock fragment sizes large enough to prevent erosion of the stone mulch itself.

(c) From a review of the literature, Tharp (1984) concluded that for natural river channels a bed is stable if D_{85} is immobile. If we apply this principle to the rill channels on the experimental plot, which have a D_{85} of 4 cm, we can conclude from Table 1 that most of the rill beds were unstable during the recorded moderate rainfall event. Thus it can be stated that rill channel armouring will be overcome several times per year on the experimental plot. The formation of an armour layer in rills formed in a highly gravelled soil, as described by Foster (1982, p. 338), may therefore be limited to low-magnitude rainfall events.

(d) Our field results also have some implications for archaeology. Vermeersch (1989) states that in the Belgian loam region, mesolithic sites are virtually unknown. Furthermore, this author assumes that if mesolithic sites were located on loam slopes, they have by now disappeared due to soil erosion. Our findings give more insight into the processes which are responsible for the evacuation of artifacts from loam-covered slopes; i.e. rill flow can easily transport flint artifacts with intermediate diameters up to 9 cm (Table 1) during moderate rainfall events, while the competence of inter-rill flow during such events is almost one order of magnitude smaller. Since rilling is very likely to occur on bare loam covered slopes having slope angles greater than 2 to 3° (Savat & De Ploey, 1982), and since forest clearance and cultivation of the soils in the Belgian loam belt was initiated at least since medieval times, the probability of downslope transport of artifacts by rill flow in the loam region is very high.

(e) From the field observations it can safely be said that rill and/or gully flows are the main processes responsible for the evacuation of rock fragments from cultivated upland areas in the Belgian loam region. Hence, the presence of rock fragments in colluvial deposits (Fig. 9) is an important indicator of the type of processes acting on the upland areas during colluviation. From Fig. 9, it can be concluded that recent colluvial deposits contain more rock fragments than older, historical colluvial deposits. This can be attributed either to the fact that more and more rock fragments have become available at the upland soil surface for subsequent evacuation, or to an increased frequency of rilling. The latter could then be an indication of accelerated soil erosion in the area.

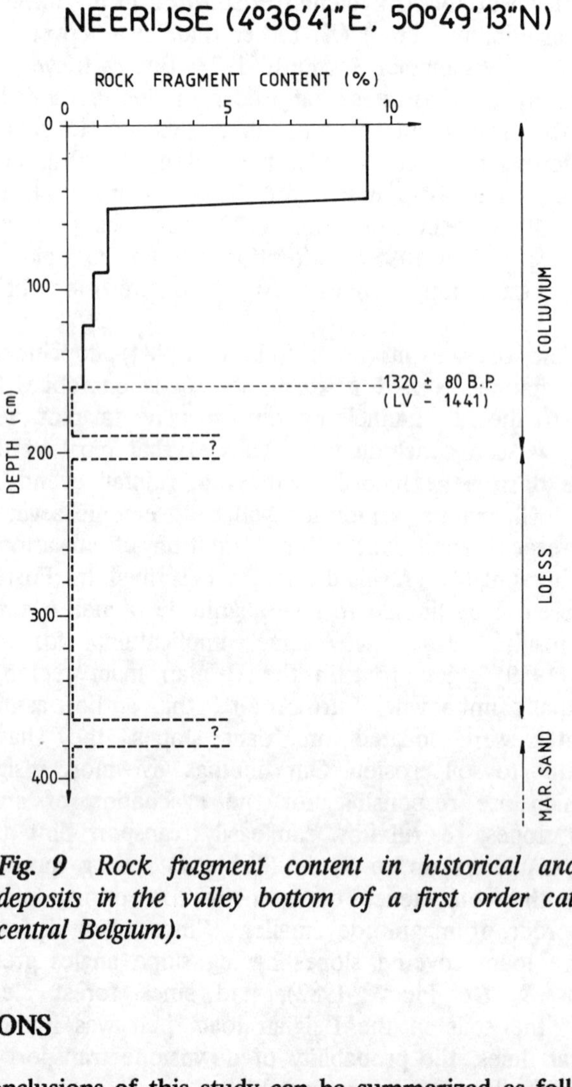

Fig. 9 Rock fragment content in historical and recent colluvial deposits in the valley bottom of a first order catchment (Neerijse, central Belgium).

CONCLUSIONS

The main conclusions of this study can be summarized as follows:

(a) Rill flow, generated during moderate rainfall events, can be identified as the most important process leading to the downslope movement of rock fragments on upland areas in the Belgian loam region. The competence of rill flow during such events exceeds by almost a factor 10 the competence of inter-rill flow. Obviously, during extreme rainfall events, rill flow will be even more effective in moving rock fragments.

(b) Incipient motion conditions, for single or clustered rock fragments lying on a relatively smooth field rill bed, coincide with a mean critical

Shields entrainment parameter (θ_c) of 0.012. This θ_c value is smaller than the generally accepted θ_c value for conventional open channel flows having gentle slopes, large ratios of flow depth to sediment size, and fine bed materials (i.e. $\theta_c = 0.06$). Laboratory data suggest that the low θ_c values corresponding to the field rills can be explained by the low rill bed roughness and, hence, a corresponding low angle of repose for the rock fragments lying on such a bed.

(c) With respect to the influence of rock fragment properties upon the distance over which rock fragments were moved by rill flow, it can be stated that fragment size plays a more important role than fragment shape. Nevertheless, for a given fragment diameter a large variation in transport distance is observed. Furthermore, it was shown that rock fragment displacement distances always correlate better with rill bed slope than with peak rill flow unit discharge and, hence also with rill catchment area. The critical slope gradient for rill flow transport, of 1 and 4 cm diameter rock fragments, varied between 2° 20' and 3° 30'.

Acknowledgements Dr G. Govers, Dr G. Rauws and Mr L. Cleeren are thanked for their assistance in the field. I also wish to thank Dr G. Wyseure who placed the rainfall records at my disposal. Mr R. Geeraerts is thanked for drawing the illustrations.

REFERENCES

Abrahams, A., Luk, S. & Parsons, A. (1988) Threshold relations for the transport of sediment by overland flow on desert hillslopes. *Earth Surf. Proc. Landforms* **13**, 407-419.

Adams, J. (1966) Influences of mulches on runoff, erosion and soil moisture depletion. *Soil Sci. Soc. Am. Proc.* **30**, 110-114.

Andrews, E. (1983) Entrainment of gravel from naturally sorted riverbed material. *Geol. Soc. Am. Bull.* **94**, 1225-1231.

Box, J. & Meyer, L. (1984) Adjustment of the Universal Soil Loss Equation for cropland soils containing coarse fragments. *SSSA Spec. Publ.* **13**, 83-90.

Collinet, J. & Valentin, C. (1984) Evaluation of factors influencing water erosion in West Africa using rainfall simulation. In: *Challenges in African Hydrology and Water Resources* (Proc. Harare Symposium, July, 1984), IAHS Publ. no. 144, 451-461.

Deffontaines, J. & De Montard, F. (1968) Essai d'appréciation du danger d'érosion dans les vergers de fortes pentes en moyen Vivarais. *Ann. Agron.* **19**, 349-364.

De Wilde, L. (1986) Studie van de competentie van stroming in geulen in functie van helling, debiet en beddingsruwheid. Unpublished MSc thesis, K. U. Leuven.

De Ploey, J. & Moeyersons, J. (1975) Runoff creep of coarse debris: experimental data and some field observations. *Catena* **2**, 275-288.

Epstein, E., Grant, W. & Struchtemeyer, R. (1966) Effects of stones on runoff, erosion and soil moisture. *Soil Sci. Soc. Am. Proc.* **30**, 638-640.

Fenton, J. & Abbott, J. (1977) Initial movement of grains on a stream bed: the effect of relative protrusion. *Proc. Roy. Soc. Lond.* **352**, 523-537.

Foster, G. (1982) Modeling the erosion process. In: *Hydrologic Modelling of Small Watersheds* (ed. Haan, C.) Am. Soc. Agric. Engrs. Monograph **5**, 297-380.

Foster, G., Huggins, L. & Meyer, L. (1984) A laboratory study of rill hydraulics: I. Velocity relationships. *Trans. Am. Soc. Agric. Engrs.* **27**, 790-796.

Govers, G. (1986) Mechanism van akkererosie op lemige bodems. Unpublished PhD thesis, K. U. Leuven.

Govers, G. & Poesen, J. (1986) A field-scale study of surface sealing and compaction on loam and sandy loam soils. Part I. Spatial variability of soil surface sealing and crusting. In: *Assessment of Surface Sealing and Crusting* (eds. Callebaut, F., Gabriels, D. & De Boodt,

M.), R. U. Gent, 171-182.

Jennings, G. & Jarrett, A. (1985) Laboratory evaluation of mulches in reducing erosion. *Trans. Am. Soc. Agric. Engrs* **28**, 1466-1470.

Kochenderfer, J. & Helvey, J. (1987) Using gravel to reduce soil losses from minimum-standard forest roads. *J. Soil Wat. Conserv.* **42**, 46-50.

Laurant, A. (1976) Nouvelles recherches sur les intensités maximums de précipitation à Uccle. Courbes d'intensité-durée-fréquence. *Ann. Travaux Publics Belgique* **4**, 320-328.

Leopold, L., Emmett, W. & Myrick, M. (1966) Channel and hillslope processes in a semiarid area in New Mexico. *USGS Prof. Pap.* 352-G.

Meeuwig, R. (1970) Sheet erosion on intermountain summer ranges. *USDA Forest Service Res. Pap.* INT-85.

Meyer, L., Johnson, C. & Foster, G. (1972) Stone and woodchip mulches for erosion control on construction sites. *J. Soil Wat. Conserv.* **27**, 264-269.

Miller, F. & Guthrie, R. (1984) Classification and distribution of soils containing rock fragments in the United States. *SSSA Spec. Publ.* **13**, 1-6.

Moeyersons, J. & De Ploey, J. (1976) Quantitative data on splash erosion, simulated on unvegetated slopes. *Z. Geomorph. Suppl.* **25**, 120-131.

Newson, M. (1980) The erosion of drainage ditches and its effect on bedload yields in mid-Wales: reconnaissance case studies. *Earth Surf. Processes* **5**, 275-290.

Poesen, J. & Govers, G. (1986) A field-scale study of surface sealing and compaction on loam and sandy loam soils. Part II. Impact of soil surface sealing and compaction on water erosion processes. In: *Assessment of Soil Surface Sealing and Crusting* (eds. Callebaut, F., Gabriels, D. & De Boodt, M.), R. U. Gent, 183-193.

Savat, J. & De Ploey, J. (1982) Sheetwash and rill development by surface flow. In: *Badland Geomorphology and Piping* (eds. Bryan, R. & Yair, A.), Geo Books, Norwich, 113-126.

Schick, A., Lekach, J. & Hassan, M. (1987) Bed load transport in desert floods: observations in the Negev. In: *Sediment Transport in Gravel Bed Rivers* (eds. Thorne, C., Bathurst, J. & Hey, R.), Wiley, 617-642.

Seginer, I., Morin, J. & Shachori, A. (1962) Runoff and erosion studies in a mountainous terra-rossa region in Israel. *Bull. IASH,* 7, 79-92.

Tharp, T. (1984) Sediment characteristics and stream competence in ephemeral intermittent streams, Fairborn, Ohio. *Catena Suppl.* **5**, 121-136.

Vanoni, V. (1977) Sedimentation Engineering. *ASCE Manuals and Reports on Engineering Practice* no. 54.

Vermeersch, P. M. (1989) Ten years' research on the mesolithic of the Belgian lowland: results and prospects. In: *The Mesolithic in Europe,* papers presented at the Third Int. Symp. on the Mesolithic in Europe, Edinburgh, 1985, ed. C. Bonsall, pub. John Donald, Edinburgh, UK, 284-290.

Wischmeier, W. & Smith, D. (1978) Predicting rainfall erosion losses. *USDA Agric. Handbook* no. *537.*

Erosion, Transport and Deposition Processes (Proceedings of the Jerusalem Workshop, March-April 1987). IAHS Publ. no. 189, 1990.

Seasonal variations of runoff rates from field plots in the Federal Republic of Germany and in Hungary during dry years

G. RICHTER

University of Trier, Geographie/Geowissenschaften, Postfach 3825, D-5500 Trier, Federal Republic of Germany

A. KERTÉSZ

Hungarian Academy of Sciences, Postafiok 64, H-1388 Budapest, Hungary

Abstract Plot measurements from the Mosel region (FRG) and from northern Hungary have been used to compare seasonal variations of surface runoff and sediment production in two different types of moist-temperate climate. In addition to seasonal contrasts between the summer and winter periods, the following differences have been detected: under the maritime west coast climate of the western part of the Federal Republic of Germany, surface runoff and sediment production during the summer period exceed that in the winter; in contrast, under the humid continental climate of Hungary, having a mediterranean influence, autumn rainfall and snowmelt in early spring often exert a dominant role. This situation occurred during the measurement period. There are also years with a summer maximum in runoff. Besides precipitation distribution, temperature and soil moisture regimes contribute to the above findings obtained under extreme conditions. The measurement series from both areas point to the existence of semiarid years with 300–450 mm precipitation per year, which are prone to further modifications.

INTRODUCTION

In general the temperate climate of Central Europe is humid during all months of the year. Walter diagrams show that Germany never experiences aridity, and that Hungary only approaches semiarid conditions in September (Fig. 1). Where the curves of temperature and precipitation cross during the summer, as in the case of Haifa, Israel, semiarid or arid months are registered.

There are, however, years with long lasting dry summer periods even in Central Europe. Two examples of such dry years will be discussed, one associated with the maritime temperature climate of western Germany and the other with the continental temperature climate of Hungary. By using plot-measurements, runoff and soil loss during such dry years will be considered.

The Mertesdorf experimental station (Trier) in the Federal Republic of

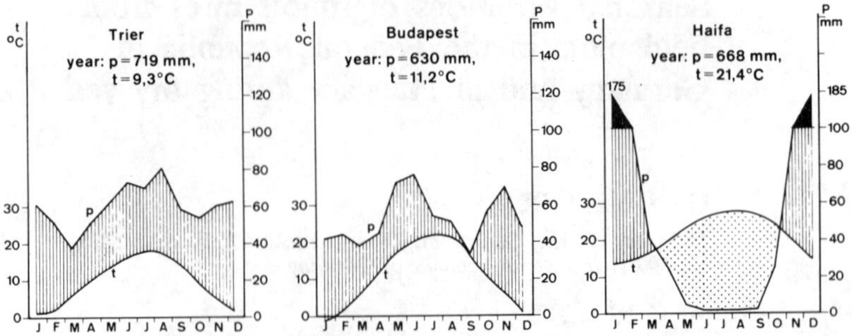

Fig. 1 Walter diagrams for the meteorological stations at Trier, Budapest and Haifa (t = temperature, p = precipitation).

Germany lies on the steep slopes of the Ruwer valley, a tributary of the Mosel. The slopes are covered with vineyards. The plots are 16 × 2.60 m (41.6 m^2) and have a slope angle of 25°. The rigosols developed on weathered slate have a profile depth of about 1 m, and a *K*-value of about 0.2. The mean annual precipitation is 720 mm.

The Pilismarót experimental station in Hungary lies near the well known "bend of the Danube" north of Budapest. The plots are 8 × 1 m (8 m^2) and have a slope angle of 8°. The *K*-value of the colluvial brown earth soils developed on loess is about 0.4. The mean annual precipitation of Budapest is 630 mm.

RESULTS

Although the values of soil loss from both sets of plots were adjusted to represent plots of 8° and 50 m, by use of the metric LS-diagram of Wischmeier, the soil losses cannot be directly compared. The main aim of this paper is, however, to compare the seasonal variation of runoff and soil loss during years with dry summer periods, such as occurred at Mertesdorf in 1975 and 1976 and at Pilismarót in 1984.

The characteristics of the rainfall distribution over the year are similar (Fig. 2): both stations have a minimum in late winter and early spring, a summer maximum and a second maximum in late autumn. A relatively dry period in Budapest during the late summer, however, reflects the mediterranean and continental influence.

With annual precipitation totals of 538 mm and 335 mm (the norm is 719 mm), the years of 1975 and 1976 were relatively dry in Trier. The Walter diagram shows dry conditions lasting from April to August 1976 (Fig. 3). The higher rainfall during July was caused by thunderstorms.

Similar conditions existed in Pilismarót in 1984 when the annual rainfall was 411 mm as compared to the norm of about 630 mm. The Walter diagram shows two dry months. In reality there were three, because the rainfall of August was nearly all associated with a single thunderstorm (Fig. 3). A neighbouring station a few kilometers away registered only 5 mm

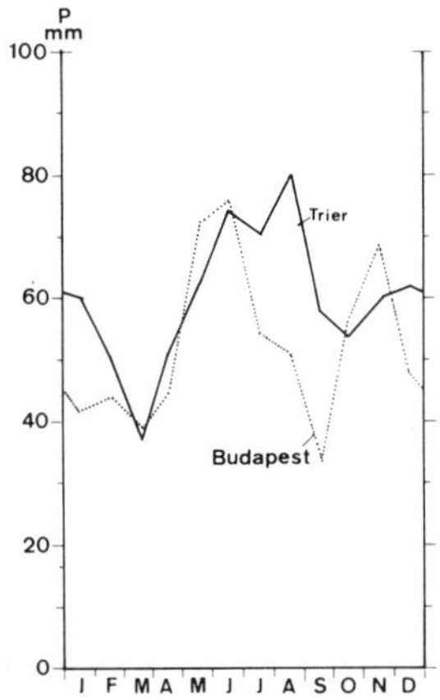

Fig. 2 The characteristic annual distribution of precipitation for Budapest (mean annual precipitation = 630 mm) and Trier (mean annual precipitation = 719 mm).

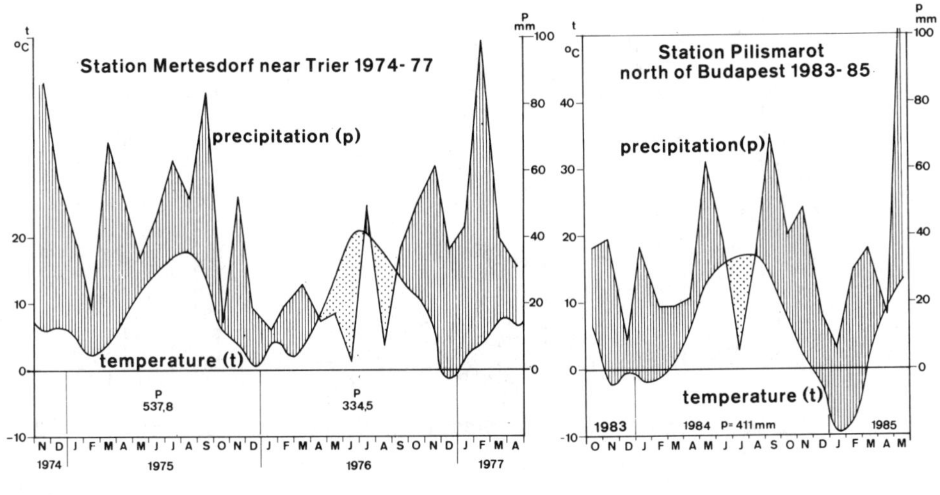

Fig. 3 Walter diagrams for dry years in Central Europe.

of rain as a total for the month.

Figure 4 shows the values of monthly runoff and soil loss for Mertesdorf. Runoff was high during the two winter seasons, for there were only short periods (1–2 weeks) with snowfall and snowmelt, which were interrupted by periods with rain. However, these winter seasons showed no soil loss. Towards spring the runoff values diminished in response to the lower precipitation and increased temperatures.

The two summer seasons were quite different. 1975 was a summer with several thunderstorms. They caused some high runoff events and soil losses. 1976 had a dry spring and summer, and there were no heavy thunderstorms. As a result it was a year with very low runoff and no soil loss.

Figure 5 shows the values of monthly rainfall, runoff and soil loss for

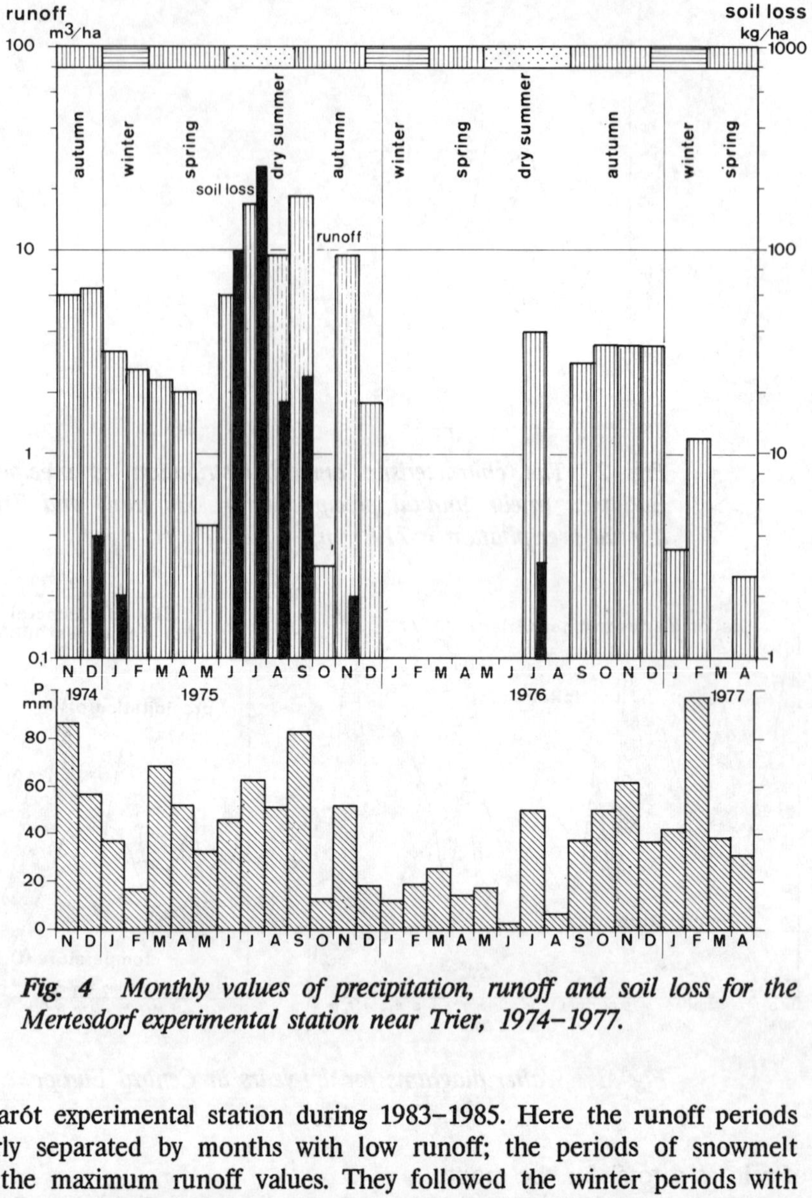

Fig. 4 Monthly values of precipitation, runoff and soil loss for the Mertesdorf experimental station near Trier, 1974–1977.

the Pilismarót experimental station during 1983–1985. Here the runoff periods were clearly separated by months with low runoff; the periods of snowmelt produced the maximum runoff values. They followed the winter periods with snow cover and without soil loss. If rainfall occurred during the melt period, as in March 1985, moderate soil losses occurred. The rainfall during the spring and the autumn caused runoff, but only the autumn showed moderate soil loss, a consequence of higher rainfall intensities. The dry summer of 1984 was generally a season with very low runoff, but in August a single thunderstorm caused high runoff and soil loss of nearly 2 t ha^{-1}.

DISCUSSION

This presentation of the data, based on monthly values, provides only a

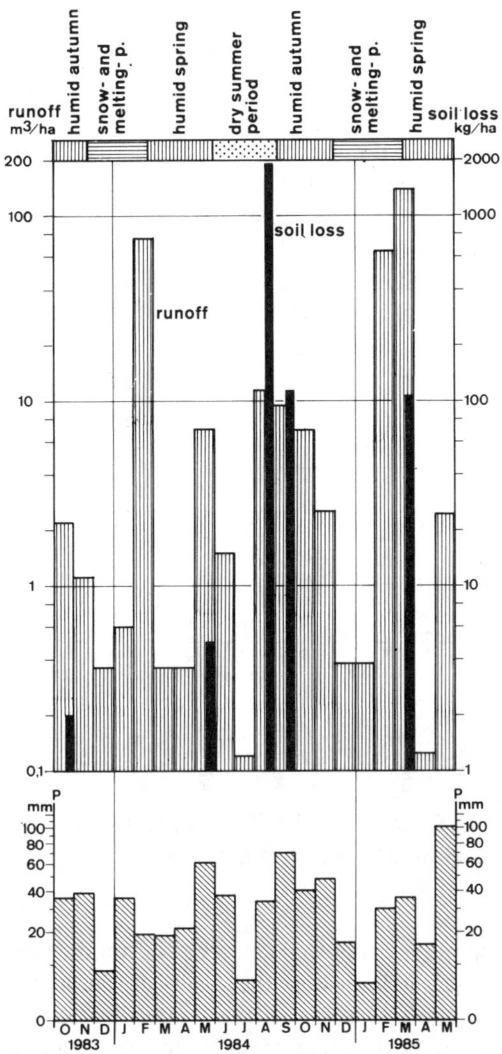

Fig. 5 Monthly values of precipitation, runoff and soil loss for the Pilismarót experimental station north of Budapest, 1983–1985.

limited view of the situation. The values were therefore aggregated for the natural seasons during the measurement period.

For the Mertesdorf station, the criteria for delimiting the natural seasons were:

winter: daily mean temperature below 5°C, short frost periods with snow and snowmelt, cyclonic rainfall less than in autumn and with low intensity;

spring: daily mean temperature 5–15°C, cyclonic rainfall of small amount and low intensity, sometimes short snow and sleet showers;

summer: daily mean temperature above 15°C, cyclonic and convective rainfall, the latter sometimes with high amounts and intensities, some dry periods;

autumn: daily mean temperatures 15–5°C, no frost days, mainly cyclonic but some convective rainfall of low to medium intensity. Towards the winter the amount of rainfall increases with an associated decrease in intensity.

For the station at Pilismarót the delimitation of natural seasons was much easier: the humid periods during the spring and the autumn were bounded by the dry summer periods and by the frost and snow periods of the winter, which ended with the period of snowmelt.

In Figs. 6 and 7 the values of runoff and soil loss for both stations are given for the natural seasons. In Mertesdorf the highest value of runoff

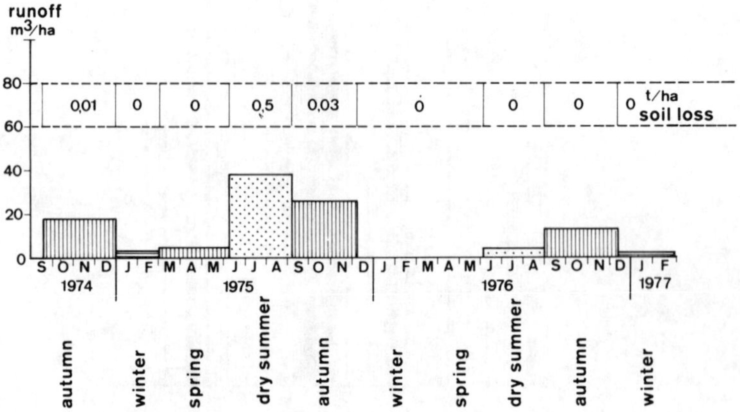

Fig. 6 Runoff and soil loss during the natural seasons 1974–1977 for the Mertesdorf station near Trier.

occurred during the summer, followed by the autumn. Only the summer season showed significant soil loss, and then only if thunderstorms occurred. At Pilismarót the main runoff season was the winter, especially during periods of snowmelt. The summer with some thunderstorms brought the maximum soil loss, although the runoff was much lower.

These features can be better expressed by diagrams showing proportional rates of runoff and soil loss during the natural seasons as a percentage of the annual total (Figs. 8 and 9). At Mertesdorf about 50% of the runoff and 90% of the soil loss occurred during the summer. At Pilismarót 60–80% of the runoff was measured during the winter season, but soil loss showed a concentration of 90% during the summer. In both regions the number of heavy thunderstorms during the dry "arid" summer period is responsible for the soil loss balance of the whole year. These significant differences in runoff can be related to the different winter climates. In the maritime temperate climate of Trier there is normally no continuous frost period during the winter. A large proportion of the rain infiltrates. As a result the runoff rate remains below 1% of the rainfall during winter and spring (Table 1). In the continental temperate climate of Hungary the soil profile was sealed by frost during

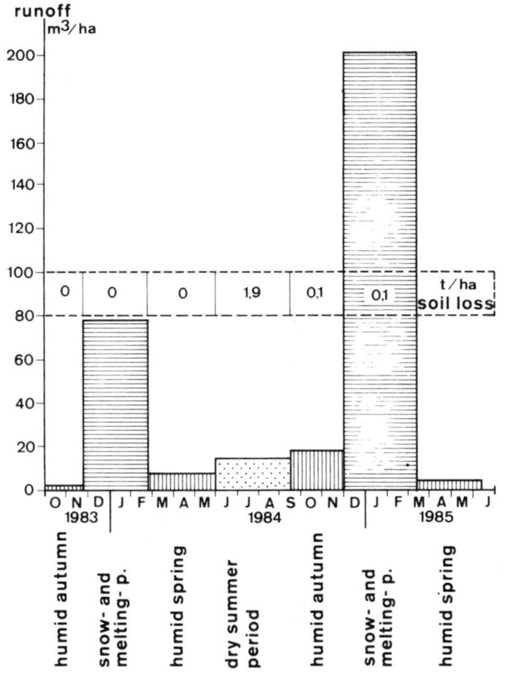

Fig. 7 Runoff and soil during the natural seasons 1983-1985 for the Pilismarót station north of Budapest.

both winter seasons. The precipitation accumulated as snow cover, and during the melt period high runoff values occurred. The runoff rate during the winter season was up to ~ 8% (winter 1983–1984) and ~ 27% (winter 1984–1985) of the precipitation (Table 1).

The dry summer periods, however, showed the same characteristics in both areas. Small rainfall events produced no runoff, but relatively high runoff rates and soil losses occurred if there were a few thunderstorms

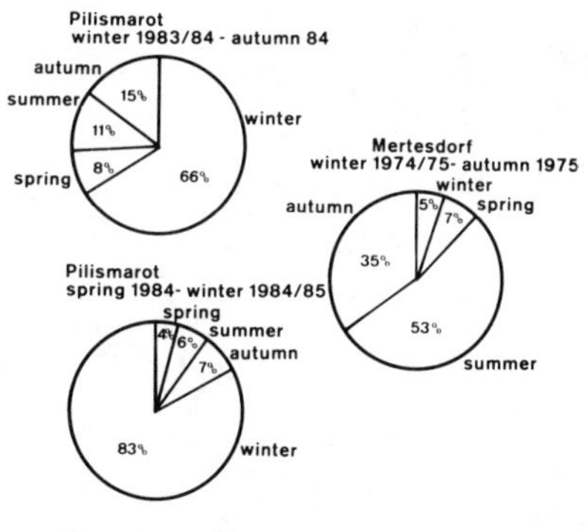

Fig. 8 Proportional runoff rates for the natural seasons.

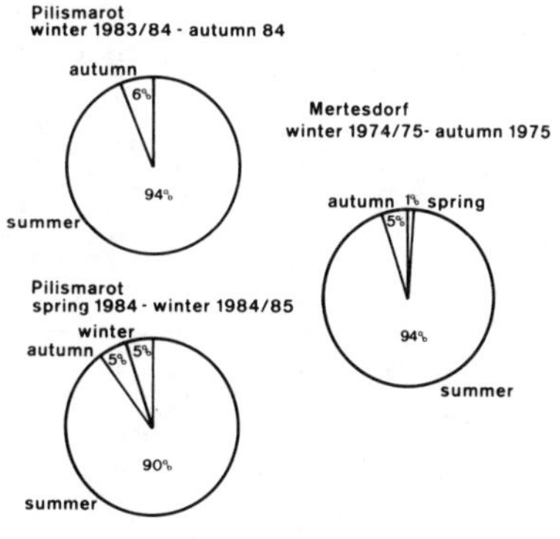

Fig. 9 Proportional rates of soil loss for the natural seasons.

Table 1 Values of precipitation, runoff, runoff rate and soil loss for the natural seasons (P = precipitation, R = runoff, R-rate = R in % of P, E = soil loss)

Station	Period	Season	P (mm)	R ($l\ ha^{-1}$)	R-rate (%)	E ($kg\ ha^{-1}$)
Mertesdorf	14.9.74 – 30.12.74	autumn	286.4	18798	0.6	12
	31.12.74 – 28.2.75	winter	65.4	3503	0.5	3
	1.3.75 – 5.6.75	spring	158.8	4800	0.3	4
	6.6.75 – 4.9.75	summer	170.6	37998	2.2	532
	5.9.75 – 3.12.75	autumn	149.1	25091	1.7	30
	4.12.75 – 23.3.76	winter	51.1	39	0	0
	24.3.76 – 5.6.76	spring	43.2	0	0	0
	6.6.76 – 1.9.76	summer	57.3	4011	0.7	4
	2.9.76 – 8.12.76	autumn	189.7	13693	0.7	3
	9.12.76 – 1.3.77	winter	143.9	1673	0.1	0
Pilismarót	29.9.83 – 25.11.83	autumn	37.6	2489	0.7	2
	26.11.83 – 25.2.84	winter	99.6	78152	7.9	0
	26.2.84 – 9.6.84	spring	128.2	9000	0.7	5
	10.6.84 – 14.9.84	summer	55.2	13500	2.5	1864
	15.9.84 – 2.12.84	autumn	159.2	18125	1.1	111
	3.12.84 – 17.3.85	winter	76.0	201625	26.5	104
	18.3.85 – 11.6.85	spring	145.4	2500	0.2	0

with high rainfall intensity. In this way, runoff and erosion react during these dry summer periods in a similar fashion to a semiarid environment.

Erosion, Transport and Deposition Processes (Proceedings of the Jerusalem Workshop, March-April 1987). IAHS Publ. no. 189, 1990.

New developments in measuring bed load by the magnetic tracer technique

RAYMUND SPIEKER & PETER ERGENZINGER

Freie Universität Berlin, Institut für Physische Geographie, Grunewaldstrasse 35, 1000 Berlin 41, FR Germany

Abstract The parallel coil magnet tracer technique is a new method for *in situ* monitoring of bed load transport. The method requires part of the bed load to be magnetized. Many drainage basins which contain igneous or metamorphic rocks will provide enough bed load material of adequate natural magnetic field intensities such that the method can be used in many regions of the world. The novel design of the detector system, in conjunction with high sensitivity data acquisition and a fast recording system, offers features such as pebble street detection, determination of the actual pebble velocity and estimation of the size of the particles in motion.

INTRODUCTION

Measuring coarse grained bed load transport under natural conditions remains an unsolved problem. The accuracy of bed load measurements when using classical trapping systems, such as the Helley-Smith sampler, is greatly affected by the extreme temporal and lateral transport rate variations that occur naturally, even during steady flow conditions (Hubbel & Stevens, 1986).

Other alternatives, for example conveyor belt and vortex bed load traps, involve high expenses and interrupt the transport process at the measuring site.

In order to improve our understanding of bed load transport and to obtain a universal model of the transport processes it is necessary to develop measuring systems which monitor the moving material with high temporal and spatial resolution without disturbing the particle interaction or the streamflow.

PREVIOUS INVESTIGATIONS

Since 1980 our working group has been engaged in developing magnetic techniques for measuring bed load movement. Various experiments using artificially magnetic tracer particles have been undertaken on rivers in Calabria (Italy) and Bavaria (FRG). Comparable investigations have been carried out by Reid *et al.* (1984) in England and by Hassan *et al.* (1984) in Israel.

In 1981 the measuring station at Squaw Creek, Gallantin County, Montana, USA, was installed in collaboration with the Montana State University. Squaw Creek drains an area which contains about 55% andesitic volcanic and intrusive rocks with a magnetite concentration of up to 10%.

An inductive detector system was installed to record the movement of these naturally magnetized pebbles (Custer *et al.*, 1987). Although initial results have been encouraging, there are some serious inherent problems which are difficult to overcome.

The magnetic field intensity of the pebbles, which is the vector sum of remanent and induced magnetism (by the earth's magnetic field), is very small. An analysis of 441 pebbles from the river bed of Squaw Creek, carried out by Monahan & O'Rourke (1982), gives the following distribution:

Field intensity in gamma (γ) units:	< 10	10-39	40-99	100-239	> 240
Number of pebbles:	261	106	38	23	13

By comparison, the field intensity of the earth's magnetic field is about 50 000 γ (1γ = 1 ntesla = 10^{-9} V s m^{-2}).

Because the induced voltage is proportional to the temporal variation of the magnetic flux in the detector system, even weak vibrations of the detectors cause electrical pulses. Other sources of noise are line noise, car ignition and lightning.

With this first measuring system, only pebbles with a field intensity greater than 240 γ induced signals which could be clearly distinguished from the background noise.

MEASURING SCHEME OF THE NEW SYSTEM

In 1986 an improved measuring method was developed and installed at Squaw Creek, with the following objectives:

(a) improvement of sensitivity of the system to obtain signals from moving pebbles even with low field intensities;
(b) reduction of background noise;
(c) detection of moving pebbles over the total cross section of the river;
(d) detection of preferred transport paths;
(e) determination of the actual velocity of the pebbles being transported; and
(f) estimation of the size of the coarse particles that cross the detector.

To achieve these objectives a new detector system was developed. Each detector unit has a length of 1.4 m and consists of over 300 chokes (small electric coils). The axes of the chokes are perpendicular to the river bottom. The chokes of each detector are serially connected so that the total inductivity of each detector is about 21 Henry (V s A^{-1}). This design guarantees both high and equal sensitivity over the entire detector. The total detector system consists of 10 independent sensor units installed in two parallel lines in a "log" which is large enough (about 8 m) to span the whole river. The distance between the lines is 0.15 m (Figs 1 & 2).

Because the surface of the detector log is completely smooth, in contrast to the extreme roughness of the river bed, the risk of pebbles accumulating on the detector surface is very low. As the induced signal is a function of

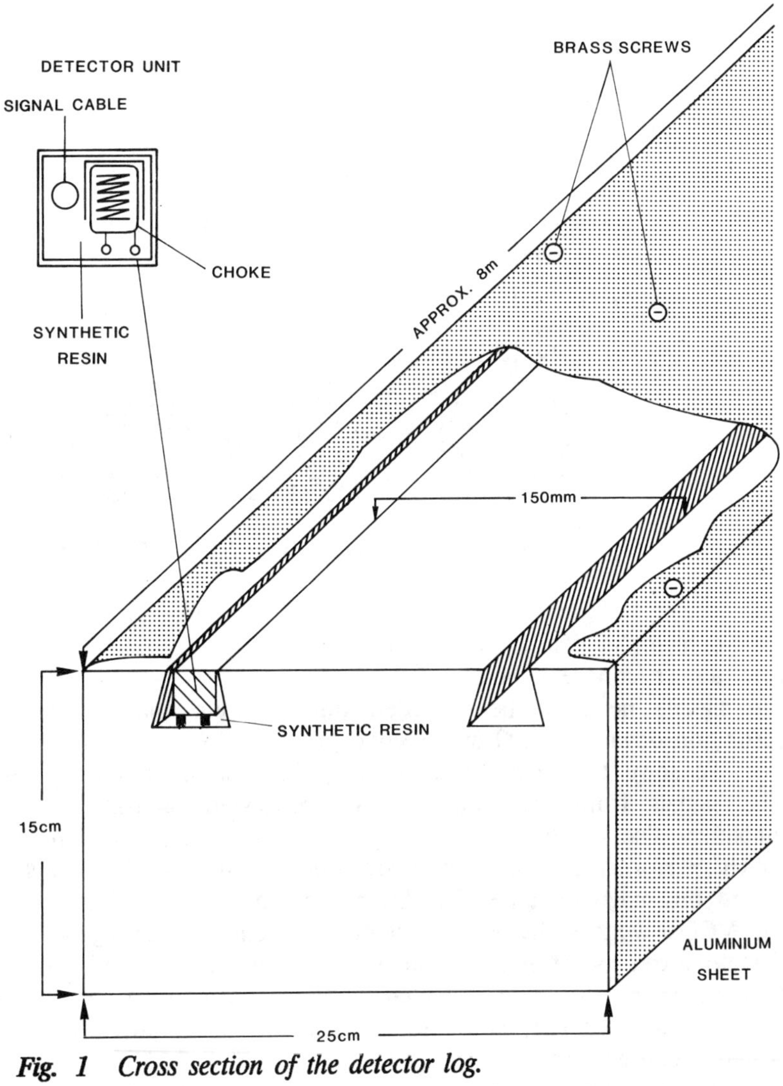

Fig. 1 Cross section of the detector log.

the velocity even the occasional settling of magnetized pebbles will not create any problems.

The arrangement of the detectors in parallel lines (parallel coil magnet tracer technique) has several advantages. External noise signals, such as log vibrations or lightning, are detected on both detector lines at the same time (coherent noise) whereas there is a distinct time different between pebble-induced signals from the upstream and downstream detector. The actual velocity of the pebble can thus be determined from the distance between the two detector lines and the time difference of the two signals.

A further problem was how to register the signals of the 10 different detector systems. In order to:

(a) obtain reliable information about the velocity of the pebbles;

(b) separate the different signals even when many pebbles cross the

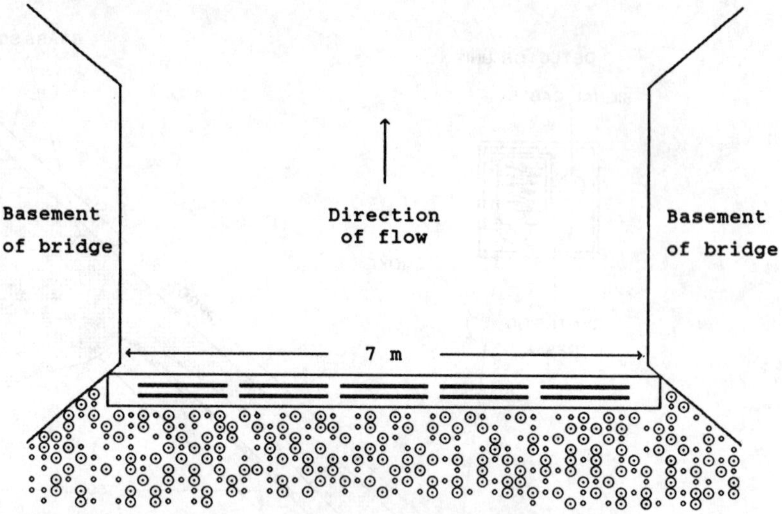

Fig. 2 View of the detector log at Squaw Creek, Montana, USA.

detector at nearly the same time; and
(c) detect coherent noise;
the signals need to be recorded on a system that permits a temporal resolution down to 1/100 of a second or less.

To solve this, a multichannel amplifier/modulator system was developed. The signals from each detector are processed separately by a high-gain amplifier. The resulting output triggers a voltage controlled oscillator (VCO) which performs a frequency modulation of the signal. This signal is then stored on magnetic tape. To obtain synchronized recording of all detectors, the VCOs work with different centre frequencies. The signals of up to four detectors can be stored on one track of the tape. In addition, a stabilized time reference signal is provided to compensate for wow and flutter. The installed tape recorder is a four channel machine allowing 16 channels to be stored synchronously.

A schematic diagram of one channel of the recording electronics and its connection to different control systems is shown in Fig. 3. In the laboratory the information on the tape is demodulated, and can be recorded on a fast chart recorder and converted into digital signals for further processing (Fig. 4).

INITIAL RESULTS

The main task of the spring 1986 field season was to install and test the reliability of the measuring system under field conditions. During the flood event from 27 May to 6 June 1986 the system functioned satisfactorily except for the first day when electronic problems were caused by an unsteady power supply.

Figure 5 shows results obtained by counting the peaks of the output from the relatively slow control chart recorder (10 mm min^{-1}). It provides a

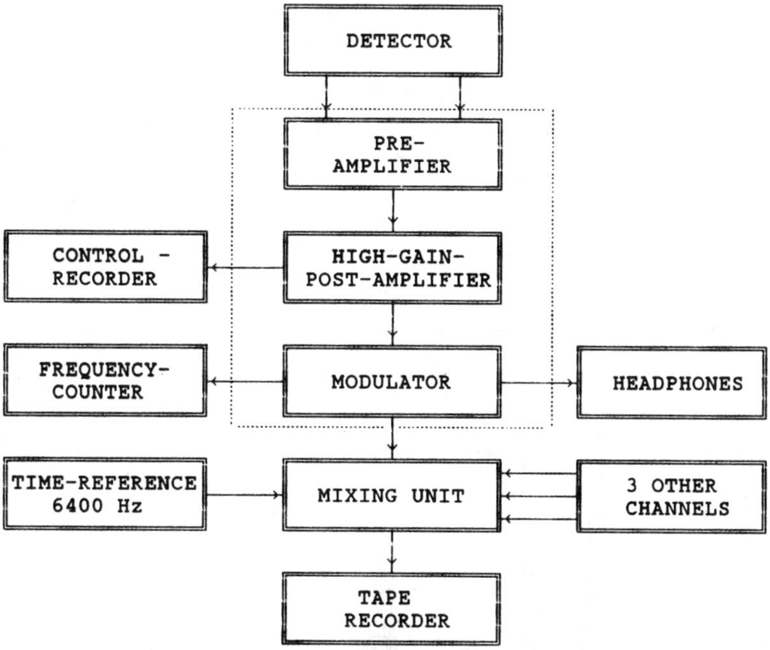

Fig. 3 Block diagram of recording electronics.

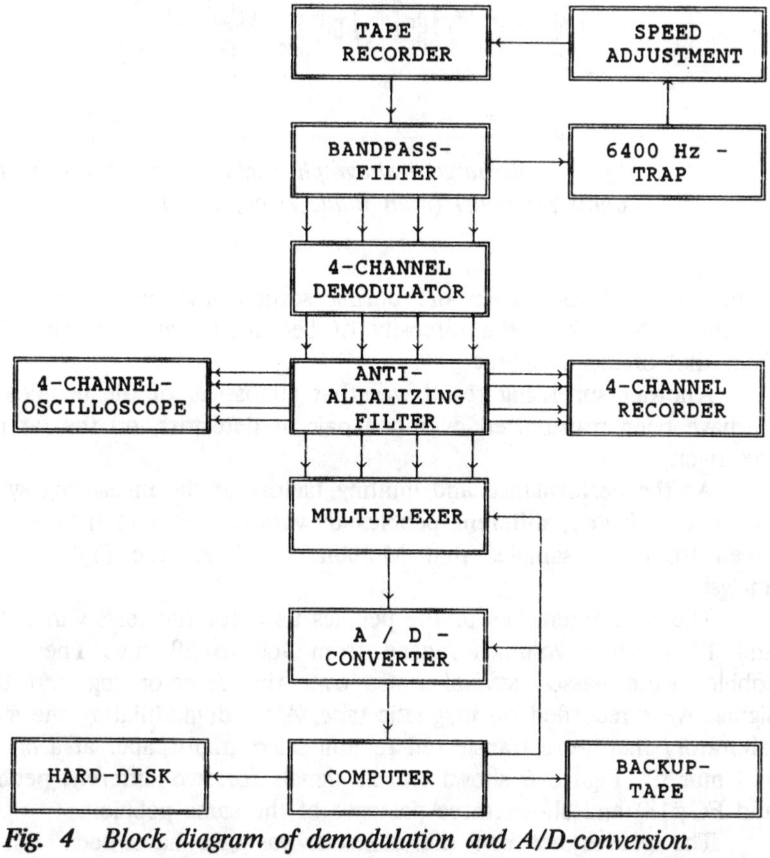

Fig. 4 Block diagram of demodulation and A/D-conversion.

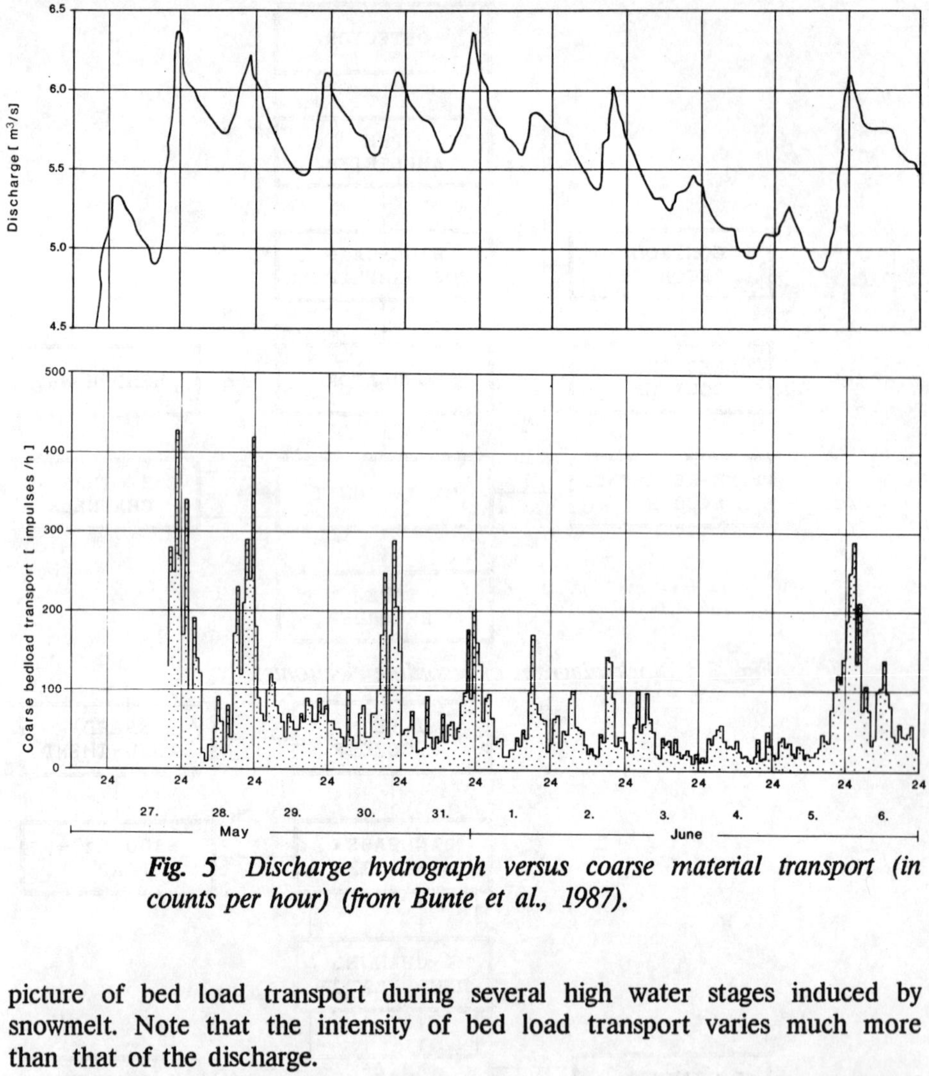

Fig. 5 Discharge hydrograph versus coarse material transport (in counts per hour) (from Bunte et al., 1987).

picture of bed load transport during several high water stages induced by snowmelt. Note that the intensity of bed load transport varies much more than that of the discharge.

Another surprising result was that almost all of the pebbles were found to have been transported over one pair of detectors, on the far right side of the river.

As the performance and limiting factors of the measuring system needed to be investigated, different pebbles of various size and field intensities were taken from the samples that Monahan & O'Rourke (1982) used in their analysis.

The field intensities of the pebbles used for the tests varied between 300 and 10 γ; their volumes ranged from 460 to 20 cm^3. These pebbles and cobbles were passed several times over the detector log and the resulting signals were recorded on magnetic tape. After demodulating the signals in the laboratory they were transferred to mm scale graph paper at a paper speed of 100 mm s^{-1}. Figure 6 shows typical signals for two different pebbles (MC/14 and FC#18) and shows three passages of the same pebble.

The pauses between the different passages have been eliminated. The upper line represents signals from the upstream detector; the lower line the signals from the parallel downstream detector. Note the differences of signal duration and amplitude when comparing axes and field intensities of the pebbles, and the time lag of the downstream signals.

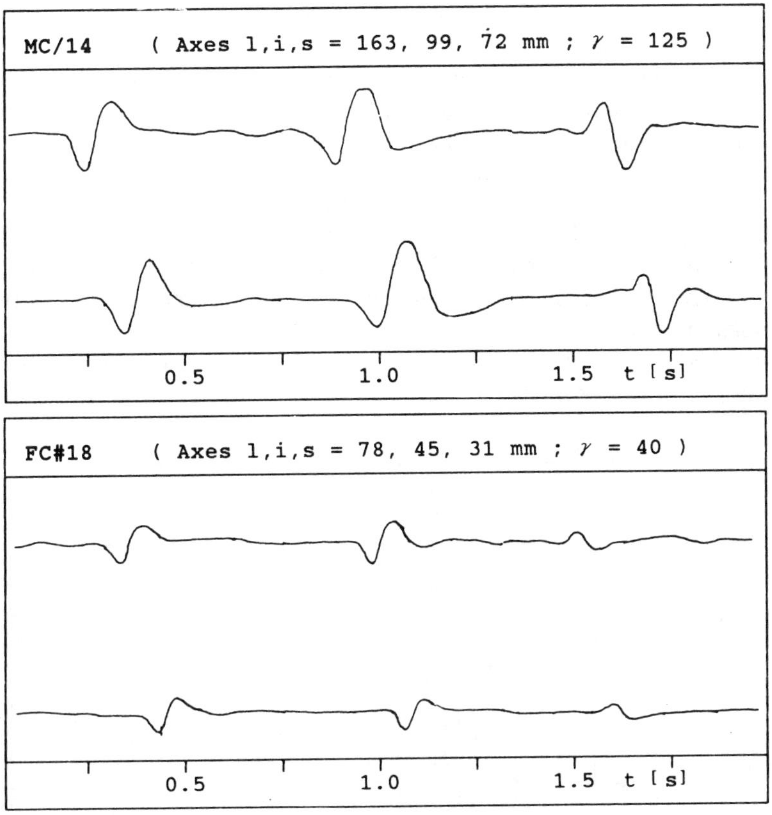

Fig. 6 Typical signals induced by naturally magnetized pebbles.

Estimation of pebble size

Previous experiments with artificial magnets showed that when the magnet is passed over the detector, a significant signal is produced in a range which corresponds to about five times the length of the projected axis of the magnet, assuming that the magnet is very close to the detector.

Similar results were obtained by computer simulations using formulae for simple geometric bodies (Telford *et al.*, 1978; Marek, 1984). However, the structure of the magnetic field of an irregularly shaped body such as a pebble, even when magnetically isotropic, is far more complicated than that of a magnetized sphere or a rod-like structure, for example. Nevertheless an attempt was made to estimate the size of the pebbles from the induced signals.

Figure 7 shows the estimated diameters of one pebble which crossed the detectors several times. The diameters were derived by multiplying the duration of the signals by the velocity of the pebble and then dividing the result by five. With a view to greater clarity the results have been sorted by size so that the x axis shows only the experiment number.

As expected, the estimated sizes vary within a certain range. This is not surprising because the orientation of the geometrical axes and especially of the axes of the magnetic vector was totally random during the experiments. Figure 7, however, also reveals that the diameters obtained by the very simple and rough estimation method are normally within the range between short and long axis of the particular pebble. The results of several experiments using different pebbles are shown in Table 1.

The table reveals that the probability of registering a pebble crossing the log increases with pebble size and field intensity. The lower limit is to be found at a volume of about 30 cm^3 and a field intensity of about 40 γ.

The minimum and maximum estimated diameters are normally in the range between short and long axes of the particular pebble. In many cases there is a surprising coincidence between the mean estimated diameters and average of the geometrical axes. Normally the deviation is less than 20%.

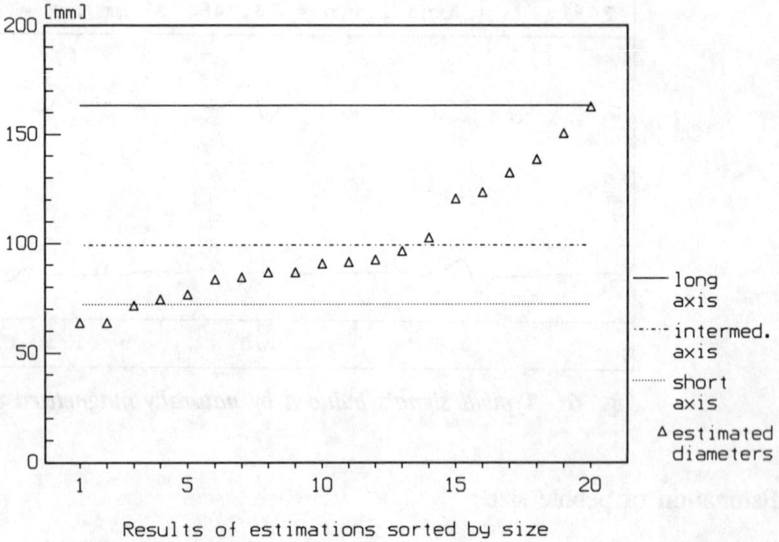

***Fig. 7** Estimation of pebble diameter: data of 20 detector passages vs real geometrical axes of pebble MC/14.*

CONCLUSION

The parallel coil magnetic tracer technique is a new method for real time recording of coarse material transport and offers new insights into bed load transport under natural conditions. It allows the detection of individual magnetized pebbles passing the measuring site. Information on pebble streets is obtained by monitoring the river cross section in several discrete steps.

Owing to the arrangement of the detectors, coherent noise signals may

***Table 1** Data of analysed pebbles*

Code	γ	Vol	Pass	Rec	L	I	S	E_{MAX}	E_{MIN}	$Ø_{LIS}$	$Ø_{EST}$
CC-24	300	120	20	20	91	78	39	93	55	69	71
MC/14	125	460	20	20	163	99	72	158	63	111	94
CC-30	125	70	25	20	82	54	36	83	36	57	57
CC-48	80	30	20	11	52	40	34	58	36	42	53
CC-21	50	40	20	12	77	58	31	61	35	55	48
CC-44	50	30	20	---	52	35	35	---	---	41	---
MC/33	40	160	20	18	111	106	35	72	33	84	50
FC#18	40	50	40	13	78	45	31	70	30	51	51
FC/29	40	30	30	16	53	38	20	52	31	37	43
FC/26	30	60	25	9	72	52	43	68	32	56	45
CC-97	30	20	20	---	55	37	30	---	---	41	---
CC-72	20	30	20	---	43	39	34	---	---	39	---
FC-14	10	80	20	4	87	44	43	59	36	58	50

Code, γ, Vol: data from Monaham & O'Rourke (1982)

Code : signature
γ : field intensity in gamma
Vol : volume of pebble
Pass : number of detector passages
Rec : recovery rate (number of registered passages)
L,I,S : long, intermediate, short axis (mm)
E_{max} : maximal estimated diameter (mm)
E_{min} : minimal estimated diameter (mm)
ϕ_{LIS} : arithmetic average of L, I and S (mm)
ϕ_{EST} : mean estimated diameter (mm)
--- : no signals received

be recognized and the actual velocity of individual particles determined. Furthermore, the size of the pebbles in motion can be estimated.

The simple size estimation method described above is not without its problems, especially when the signals of the pebbles interfere with noise signals. To overcome such difficulties, better signal analysing algorithms and error detection/correction schemes for further data processing need to be developed. For this reason a series of test runs using magnetized pebbles in a laboratory flume is planned in the near future. To facilitate wider usage, it is planned to develop programs for data acquisition and processing which run on personal computers.

REFERENCES

Bunte, K., Custer, S., Ergenzinger, P. & Spieker, R. (1987) Messung des grobgeschiebetransportes mit der magnettracertechnik. *Deutsche Gewässerkundliche Mitt.* **31**, H (2/3), 60-67.

Custer, S., Bugosh, N., Ergenzinger, P. & Anderson, B. (1987) Electromagnetic detection of pebble transport in streams. *J. Sed. Petrology* **57**, 21-26.
Hassan, M., Schick, A. & Laronne, J. (1984) The recovery of flood-dispersed coarse sediment particles. *Catena Suppl.* **5**, 153-162.
Hubbel, D. W. & Stevens, H. H. (1986) Factors affecting accuracy of bed load sampling. *Proc. 4th Federal Interagency Sedimentation Conf., Las Vegas, Nevada* **1**, 4-21 to 4-29.
Marek, F. (1984) Magnetometric methods. In: *Introduction to Applied Geophysics* (ed. Mares, S. *et al.*), 71-153. D. Reidel Publ., Dordrecht, the Netherlands.
Monahan, S. & O'Rourke, E. (1982) An analysis of the variables influencing the measurements of coarse bed load movement at Squaw Creek. Independent study at the Dept. of Earth Sciences, Montana State University.
Reid, I., Brayshaw, A. C. & Frostick, L. E. (1984) An electromagnetic device for automatic detection of bed load motion and its field applications. *Sedimentology* **31**, 269-76.
Telford, W. M., Geldart, L. P., Sheriff, R. E. & Keys, D. A. (1978) *Applied Geophysics*. Cambridge University Press, Cambridge, UK.

Erosion, Transport and Deposition Processes (Proceedings of the Jerusalem Workshop, March-April 1987). IAHS Publ. no. 189, 1990.

Some applications of caesium-137 measurements in the study of erosion, transport and deposition

D. E. WALLING & S. B. BRADLEY
Department of Geography, University of Exeter, Exeter, EX4 4RJ, UK

Abstract Caesium-137, a radionuclide originating as fallout generated by the atmospheric testing of nuclear weapons, can provide the geomorphologist with a valuable tracer for investigating the movement of sediment through the fluvial system. Three examples of possible applications drawn from the work of the authors in Devon, UK, are presented. These involve the use of ^{137}Cs for, firstly, fingerprinting suspended sediment sources, secondly, investigating patterns of soil erosion and sediment delivery on cultivated hillslopes, and thirdly, elucidating rates and patterns of flood plain deposition.

INTRODUCTION

Caesium-137 is a major component of the fallout associated with the atmospheric testing of nuclear weapons and this radionuclide is currently present in the global environment in significant quantities as a result of weapons testing during the 1950s and early 1960s. Significant fallout of ^{137}Cs was first detected in 1954 and records from a global network of measuring stations (Cambray, *et al.*, 1980; US Health and Safety Laboratory, 1977) indicate that fallout rates peaked in 1963–1964 and subsequently declined markedly as a result of the 1963 Nuclear Test Ban Treaty. Small perturbations in more recent years can be related to Chinese and French tests continuing until the late 1970s. Figure 1 provides an example of the pattern of annual fallout during the period 1950–1985 for Milford Haven, UK, based on data collected by the UK Atomic Energy Research Establishment (Cambray, personal communication). Local increases in ^{137}Cs fallout occurred in many areas of Europe and in neighbouring regions as a result of the 1986 Chernobyl accident. In some areas close to the site of the accident these inputs exceeded the accumulated fallout from weapons testing (cf. Dorr & Munnich, 1987), but in general they represented only a relatively minor contribution to the total ^{137}Cs inventory (cf. Cambray *et al.*, 1987). The total amount of weapons-testing fallout received at the land surface varies globally in response to both the magnitude of annual rainfall and the pattern of atmospheric circulation responsible for dispersing the fallout. For example, the total receipt of weapons-testing ^{137}Cs reported for sites in New South Wales, Australia by Campbell *et al.* (1986) is about 40% of that recorded for sites with similar annual rainfall in the UK.

Existing evidence indicates that in most situations, ^{137}Cs reaching the soil surface as fallout is rapidly and strongly adsorbed by clay minerals in the

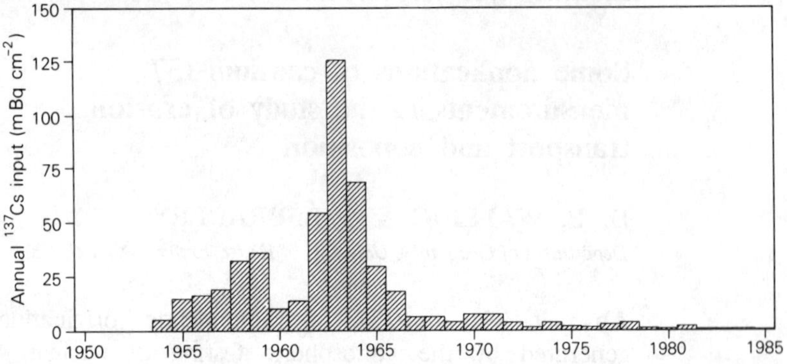

***Fig. 1** Annual fallout of ^{137}Cs recorded at Milford Haven, UK, during the period 1950–1985 (data provided by Dr R. Cambray, UK Atomic Energy Authority, Harwell).*

upper soil horizons and that further downward translocation by physico-chemical processes is limited (cf. Tamura, 1964; Gale *et al.,* 1963; Frissel & Pennders, 1983). Subsequent movement of the radioisotope is therefore generally associated with the erosion, transport and deposition of sediment particles (e.g. Rogowski & Tamura, 1970a, b; Campbell *et al.,* 1982). Because of both this behaviour and its relatively long half-life (30.1 years), which means that approximately 60% of the total input since 1954 could still remain within the system, ^{137}Cs possesses very considerable potential for use as a "natural" tracer of sediment movement. Several potential applications have been highlighted by Campbell (1983) and a recent review by Ritchie (1987) further emphasizes the range of possibilities available. Geomorphologists could profitably increase their efforts to exploit this potential and this contribution provides three examples of applications developed by the authors in Devon, UK. These involve its use for, firstly, fingerprinting suspended sediment sources, secondly, investigating patterns of soil erosion and sediment delivery on cultivated hillslopes and, thirdly, elucidating rates and patterns of flood plain deposition.

MEASUREMENT CONSIDERATIONS

Caesium-137 concentrations in soils and sediments can be measured relatively easily by gamma spectrometry. Germanium (Ge) detectors are commonly used for this purpose and the detector is generally housed in a lead shield in order to minimize background interference. The main problem with such measurements relates to the low levels of ^{137}Cs generally encountered in environmental samples. These necessitate long counting times. Counting times of the order of 36 000 s (10 hours) will frequently be required in order to obtain an acceptable precision and this in turn means that the number of samples that can be processed may be an important constraint in many studies. Minimum sample size may also prove a constraint, because the levels of activity commonly associated with environmental samples require a mass of at least

20 g to register a sufficient number of disintegrations.

FINGERPRINTING SUSPENDED SEDIMENT SOURCE

The context

Over the past few decades, recognition of the important role of suspended sediment in the fluvial transport of nutrients and contaminants and in non-point pollution from land use activities has focussed increasing attention on the sources involved and the pathways associated with the conveyance of sediment from its source to the basin outlet (e.g. Glymph, 1975; Wolman, 1977). In addition to an assessment of the magnitude of suspended sediment loads, information concerning the nature, relative importance and spatial distribution of sediment sources is therefore increasingly required. Similar requirements have also been promoted by recent interest in the establishment of sediment budgets for drainage basins (e.g. Swanson *et al.,* 1982), by advances in phsyically-based distributed modelling of sediment yield (e.g. Beasley *et al.,* 1982) and by a desire for more meaningful geomorphological interpretation of sediment yield data in terms of landscape evolution (e.g. Finlayson, 1978).

The development of the "fingerprinting" technique as a means of assessing sediment sources within a drainage basin has, therefore, attracted considerable interest, in that it offers a relatively simple means of establishing the relative importance of major sediment sources (e.g. Wall & Wilding, 1976; Wood, 1978; Oldfield *et al.,* 1979). In essence this method involves firstly, the selection of a physical or chemical property which clearly differentiates potential source materials, and, secondly, a comparison of measurements of this property obtained from suspended sediment with those for the potential sources. Additional considerations relating to the choice of appropriate sediment properties and procedures for establishing the relative importance of individual source materials are discussed by Peart & Walling (1986).

The viability of this approach to determining the relative importance of the major sediment sources within a drainage basin depends heavily upon the availability of a sediment property capable of clearly distinguishing sediment derived from different sources such as cultivated fields and channel banks. Work undertaken by the authors suggests that the ^{137}Cs concentration of source material and suspended sediment affords an excellent fingerprint. Different source materials are characterized by different concentrations of ^{137}Cs which will in turn reflect the amount of fallout received and its degree of incorporation within the soil profile. Channel banks, which are in most cases essentially vertical, will receive relatively little fallout as compared to field areas and will therefore be characterized by very low concentrations, or even an absence, of ^{137}Cs. Caesium-137 reaching the soil surface will be strongly adsorbed by the fine fraction and will therefore be preferentially concentrated near the surface. Where cultivation occurs, however, this surface concentration will be mixed throughout the depth of cultivation (ca. 10–30 cm) and ^{137}Cs concentrations in surface topsoil will be appreciably lower than those encountered in uncultivated soil. A typical example of the distribution of ^{137}Cs in the upper soil horizons of permanent pasture and of arable fields in the vicinity of Exeter, UK, is provided in Fig. 2. Here, sediment eroded

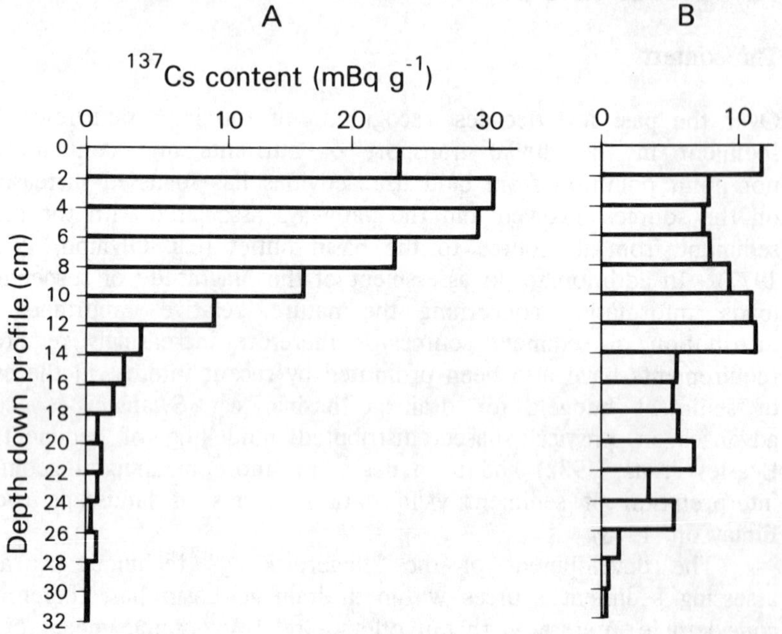

Fig. 2 The vertical distribution of ^{137}Cs in soil profiles representative of (A) undisturbed pasture and (B) arable fields within the Jackmoor Brook basin.

from cultivated fields could be expected to be characterized by ^{137}Cs concentrations that are only about 40% of those associated with sediment derived from uncultivated areas. Both sources would, however, provide sediment with ^{137}Cs concentrations about an order of magnitude greater than those encountered in sediment derived from channel bank sources. By virtue of its dependence on atmospheric fallout, which can be viewed as essentially uniform over small areas, the ^{137}Cs concentration of sediment and source material provides a fingerprint which is largely independent of soil and regolith properties which may exhibit marked spatial heterogeneity.

An example

The potential offered by ^{137}Cs as a means of fingerprinting sediment source can be further demonstrated by considering the results obtained from a study undertaken by Peart & Walling (1986) in the Jackmoor Brook basin, near Exeter, UK. This small drainage basin (Fig. 3) has a drainage area of 9.3 km^2 and its altitude ranges from 21.5 m at the basin outlet to 235 m on the

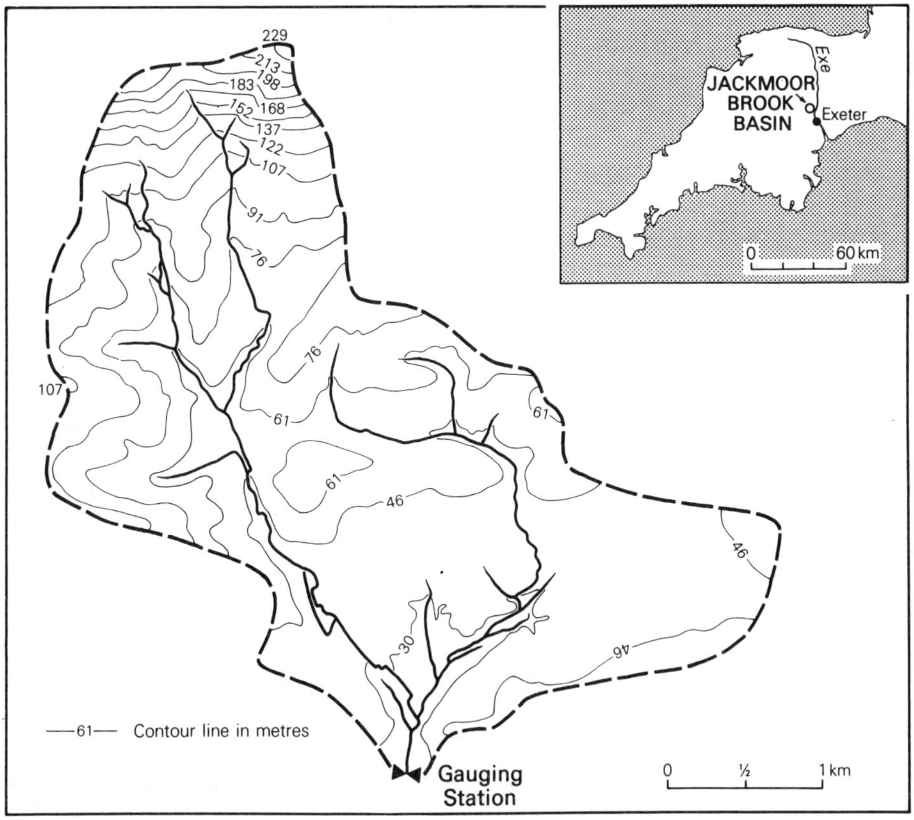

Fig. 3 The Jackmoor Brook study basin.

northern divide. Slopes are predominantly gentle (<4°), although steeper ground occurs towards the northwestern margin (8°). The basin is underlain by Permian sandstones, breccias and conglomerates. Land use is predominantly mixed arable farming, with cattle and sheep rearing and a rotation of cereals, root crops and grass. The majority of the grassland is in leys. Mean annual precipitation and runoff are estimated at 825 mm and 350 mm respectively. Within the local area, this drainage basin is noteworthy for the relatively high suspended sediment concentrations of up to 3500 mg l^{-1} which may occur during storm events. Based on available records the mean annual suspended sediment yield is estimated to be c. 60 t km^{-2} $year^{-1}$. The stream network is only moderately incised into the valley bottoms, and bank heights are commonly less than 1 m although deeper incision of 2–3 m occurs in places. Flood plain development is limited and active channel incision reaches bedrock in most locations. Neither the channel network nor the agricultural land with its pattern of hedge bank boundaries provide obvious evidence of major sediment sources and for this reason the fingerprinting technique appeared to offer a useful means of identifying the dominant sediment sources.

Three potential sediment sources were distinguished within the basin, namely, channel banks, arable fields and permanent pasture, and representative source material was collected from over 100 sites within the basin. These sites were selected to embrace the major components of variability within the three source types. Analysis of these samples was restricted to the <63 μm fraction, in order to minimize differences in particle size composition between source material and suspended sediment. Available evidence indicated that the >63 μm fraction of the suspended sediment load transported by the Jackmoor Brook rarely exceeded 5%. In the knowledge that suspended sediment properties would be likely to vary according to season and in response to variations in discharge and suspended sediment concentration (e.g. Walling & Kane, 1982), suspended sediment sampling was undertaken over a range of conditions. Bulk samples of river water (c. 100 l) were collected during storm events using a pump sampler and the suspended sediment was recovered using a continuous flow centrifuge. The suspended sediment samples were freeze dried prior to analysis.

Mean values of ^{137}Cs content for the three potential source materials and for suspended sediment are listed in Table 1 (A). The high values of

Table 1 *Comparison of the ^{137}Cs content of suspended sediment transported by the Jackmoor Brook with that of potential source material associated with arable fields, permanent pasture and bank material*

Caesium-137 content (mBq g^{-1})	*Bank material*	*Arable topsoil*	*Pasture topsoil*	*Suspended sediment*
A	*1.5*	*9.0*	*23.0*	*17.0*
B	*2.25*	*13.5*	*34.5*	*17.0*

A = Raw data
B = Data corrected for particle size enrichment

^{137}Cs content associated with suspended sediment indicate that channel banks are unlikely to be a significant sediment source in this basin and suggest that cultivated fields and pasture provide the dominant source. A direct comparison of the ^{137}Cs levels associated with suspended sediment and potential source materials is, however, likely to yield misleading results, because of the enrichment of suspended sediment in clay-sized particles relative to the source material. In view of the preferential association of ^{137}Cs with the clay fraction, suspended sediment derived from a particular source will inevitably contain higher concentrations of the radionuclide than the source material. Measurements of the grain size composition of suspended sediment and of the three potential source materials indicated that the former typically exhibited an enrichment factor for fines of the order of 1.5. The values of ^{137}Cs concentration listed for the source materials in Table 1 have therefore been increased by applying this enrichment factor of 1.5, in order to make them directly comparable with that for suspended sediment. The amended values listed in Table 1 (B) suggest that arable fields are likely to provide the dominant source of the suspended sediment transported by the Jackmoor Brook, since the fingerprint values associated with topsoil from areas of permanent pasture are about double those characterizing suspended sediment. Furthermore, permanent pasture accounts for only a small proportion of the basin area.

Although the fingerprint evidence provided in Table 1 (B) points strongly to arable fields as providing the dominant sediment source in this basin, the possibility that the values of ^{137}Cs associated with suspended sediment could result from a mixture of sediment derived from areas of permanent pasture and from channel banks cannot be ruled out. Further confirmation can, however, be obtained by considering evidence from other fingerprinting properties. Peart & Walling (1986) used a number of property ratios to distinguish sediment derived from surficial and channel sources and Table 2 provides examples of the data provided by the property ratios which were judged by those authors to be the most reliable fingerprints. All four property ratios point to the dominance of surface soil as the sediment source and Table 3 lists the estimates of the relative importance of the two sources obtained by Peart & Walling

Table 2 *Comparison of mean values of selected property ratios for suspended sediment and potential source materials from the Jackmoor Brook*

Property ratio	*Bank material*	*Surface soil*	*Suspended sediment*
Carbon/nitrogen	*7.31*	*9.83*	*9.82*
SIRM/magnetic susceptibility*	*47.74*	*15.58*	*17.87*
Dithionite iron/magnetic susceptibility	*41.50*	*7.40*	*7.92*
Manganese/dithionite iron	*0.058*	*0.116*	*0.110*

**SIRM = Saturation isothermal remanent magnetization*

Table 3 *The relative contribution of bank material and surface soil sources to the total suspended load of the Jackmoor Brook estimated using a simple mixing model applied to the data in Table 2*

Property ratio	*Relative contribution (%)*	
	Surface soils	*Bank material*
Carbon/nitrogen	*99.6*	*0.4*
SIRM/magnetic susceptibility	*93.0*	*7.0*
Dithionite iron/magnetic susceptibility	*98.5*	*1.5*
Manganese/dithionite iron	*89.0*	*11.0*
Mean	*95.0*	*5.0*

(1986) using a simple mixing model of the form:

$$C_s = \frac{P_r - P_b}{P_s - P_b} \times 100$$

where: C_s = % contribution from surficial sources;
P_r = property ratio value characteristic of suspended sediment;
P_s = property ratio value characteristic of surfical soil; and
P_b = property ratio value characteristic of bank material.

The property ratios listed in Table 2 are based on intrinsic soil properties and are unable to distinguish topsoil from arable fields and permanent pasture. However, ^{137}Cs concentrations provide a means of evaluating the relative contributions of these two potential surface sources. A mixing model similar to that outlined above has been used in conjunction with the data listed in Table 1 (B) to apportion the 95% of the total sediment yield attributed to surficial sources in Table 3 between arable fields and permanent pasture. The results are presented in Table 4. These suggest that arable fields represent the dominant sediment source in this basin and account for approximately 75% of the sediment derived from surficial sources.

Table 4 *The relative contribution of topsoil from arable fields and permanent pasture and bank material to the total suspended sediment load of the Jackmoor Brook estimated using a mizing model applied to the ^{137}Cs concentration data listed in Table 1*

Source	*Relative Contribution (%)*
Channel banks	*5*
Permanent pasture top soil	*19*
Arable topsoil	*76*

In the above example of the use of ^{137}Cs to fingerprint sediment source, attention has focussed on the average values associated with suspended sediment, and therefore the relative importance of the various sources to the long term sediment yield. More detailed work could, however, investigate both seasonal and inter- and intra-storm event variations in the relative importance of individual sources. Table 5, for example, compares the average values of ^{137}Cs content associated with suspended sediment collected during the winter (October–March) and summer (April–September) periods of 1986–1987 from three sites within the Jackmoor Brook catchment, with a view to identifying any seasonal variation in the relative importance of the various sources. At all three sites, the ^{137}Cs concentration associated with suspended sediment is higher during the winter than during the summer season. This points to an increased importance of surface sources during the winter period, which is consistent with the local cultivation practice.

Table 5 *Average values of ^{137}Cs content associated with suspended sediment collected during the winter and summer periods of 1986–1987 from three sites within the Jackmoor Brook basin*

Site	*Mean ^{137}Cs concentration ($mBq\ g^{-1}$)*	
	Winter	*Summer*
Basin outlet	*17.8*	*11.2*
Yendacott	*17.6*	*12.8*
Shute	*19.5*	*13.4*

Autumn-sown cereals are the dominant crop in this area and the surface of many of the cultivated fields is therefore unprotected by vegetation during the winter period. These fields are cultivated and sown in the autumn and an appreciable crop cover does not appear until late March.

Caesium-137 concentration can provide the geomorphologist with a valuable means of fingerprinting sediment sources and evaluating the relative importance of those sources. In many instances, however, its use will be best supported by other fingerprinting properties, in order to check the consistency of the results obtained and to permit more detailed apportionment between individual sources.

INVESTIGATING PATTERNS OF SOIL EROSION AND SEDIMENT DELIVERY ON HILLSLOPES

Background

Because the ^{137}Cs reaching the soil surface in a drainage basin is under most circumstances rapidly and strongly adsorbed within the upper soil horizons and its subsequent movement associated with the erosion, transport and deposition of sediment particles, an investigation of levels of total ^{137}Cs activity within the profile ($mBq\ cm^{-2}$) at different sites could provide valuable information on the spatial distribution of erosion and deposition. If the measured values are compared with an estimate of the original input, depletion would indicate erosion, whereas areas of deposition would be marked by enhanced levels of ^{137}Cs. Estimates of the total or baseline input are commonly obtained by measuring the total activity at undisturbed sites located on an interfluve, which are unlikely to have experienced either erosion or deposition. These have been termed input sites by Campbell *et al.* (1982) and control sites by De Jong *et al.* (1983). This approach has, for example, been used by McHenry & Ritchie (1977) and McHenry *et al.* (1978) to study patterns of erosion and deposition along slope transects within the White Clay Lake basin in Wisconsin, USA; by Longmore *et al.* (1983) to map the major areas of soil erosion and deposition within an upland drainage basin on the Darling Downs, Australia, and by Loughran *et al.* (1982) to study sediment movement within a small drainage basin in the Hunter Valley,

New South Wales, Australia. The potential application of this approach can be further demonstrated by considering the results of a study undertaken by the authors in a drainage basin near Exeter, UK.

An example

Caesium-137 profiles for four sites within the Jackmoor Brook drainage basin, which was described in the previous section (cf. Fig. 3), are illustrated in Fig. 4. The samples were collected from an area of 800 cm^2 at depth increments of 2 cm using the scraper plate technique advocated by Campbell & Loughran (personal communication). Profile A was obtained from an area of permanent pasture that was considered to provide a representative input or control site for the basin. It evidences the typical down-profile distribution of ^{137}Cs described by other workers for undisturbed sites, with the majority of the radionuclide being contained in the top 10 cm and an exponential decrease occurring below. The total activity of 254 mBq cm^{-2} recorded for this site is close to the mean value of 250 mBq cm^{-2} for five input sites located in the basin which has been used as a reference or control. This mean value was in close agreement with the findings of Cambray *et al.* (1982) who suggested that the magnitude of ^{137}Cs input over the UK was closely controlled by mean annual precipitation. The remaining three profiles illustrated in Fig. 3 exhibit a contrasting down-profile distribution of ^{137}Cs which is characteristic of cultivated soils. The mixing associated with ploughing and other cultivation incorporates the ^{137}Cs more uniformly within the profile and produces a near constant down-profile distribution. The value of total activity for profile B is less than the input or control value of 250 mBq cm^{-2} and is indicative of an eroded site. The down-profile distribution suggests a maximum cultivation depth of c. 28 cm which is consistent with the agricultural machinery used in this area. In contrast to profile B, profiles C and D evidence total ^{137}Cs activities well in excess of the reference or control value of 250 mBq cm^{-2}. These high values are indicative of depositional sites where soil eroded from upslope has been deposited and subsequently incorporated within the profile by cultivation. Profile C has a ^{137}Cs "excess" of 61%, and this value is increased to 64% for profile D. The evidence for deposition provided by the excess ^{137}Cs activity in these two profiles is further substantiated by the down-profile distribution which extends well below the depth of cultivation. A simple comparison of the ^{137}Cs distributions for these two profiles with the 28 cm cultivation depth inferred above suggests that approximately 15–20 cm of sediment has been deposited at these two sites during the past 30 years. This is consistent with the location of these sites near the foot of a slope.

The assembly of ^{137}Cs profile data is extremely time consuming both in terms of the depth-incremental field sampling involved and the substantial number of samples from each site requiring analysis for ^{137}Cs content. Its potential for studying spatial patterns of erosion and deposition is consequently limited. An alternative involves the collection of single whole-core samples from each site and measurement of the total ^{137}Cs activity (mBq cm^{-2}) of the bulk sample. Comparison of these values with the reference or control value will indicate the existence of areas of erosion and deposition and provide some indication of the relative magnitude of the rates involved. In view of the simpler sampling procedure involved and the need to analyse only a single sample from each site, it is possible to establish a relatively dense network of measuring sites.

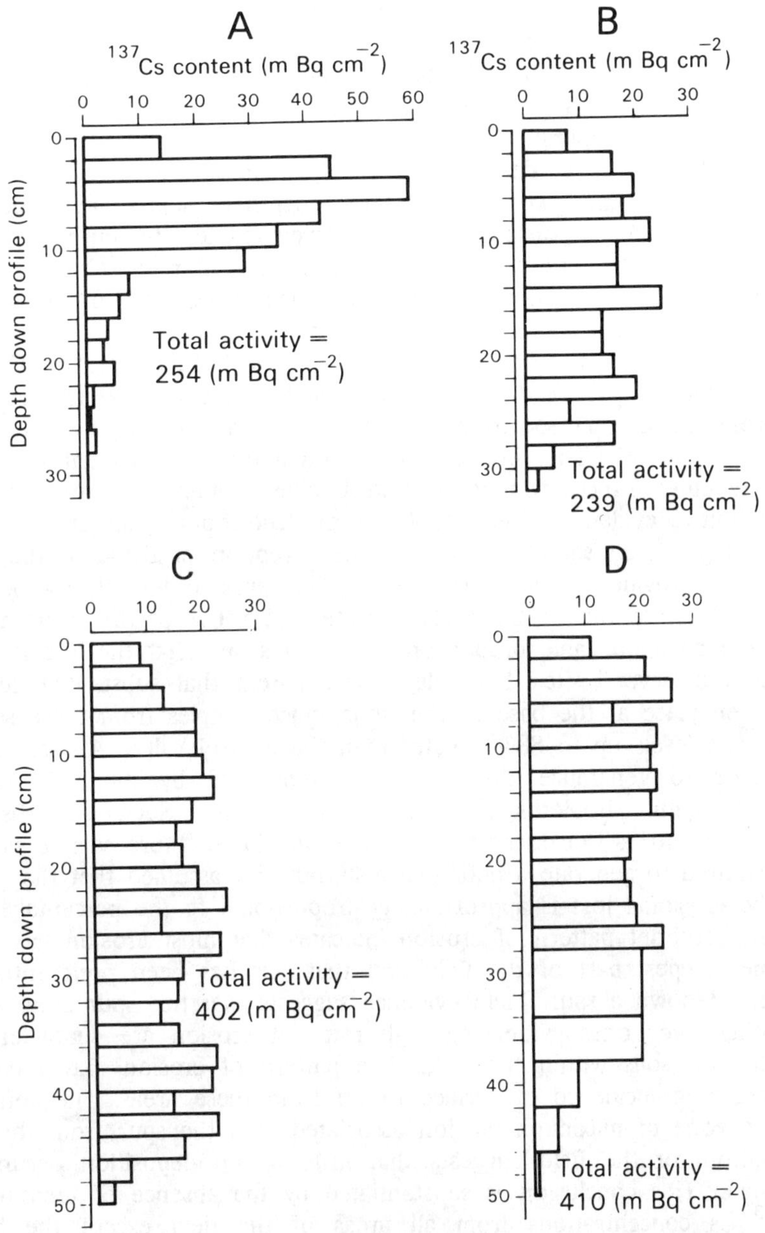

Fig. 4 Caesium-137 profiles representative of undisturbed pasture (A) and eroded (B) and depositional sites (C & D) within cultivated fields in the Jackmoor Brook basin.

This whole-core sampling approach has been used by the authors to investigate patterns of erosion and deposition on three cultivated fields, two located within, and one immediately adjacent to, the Jackmoor Brook drainage basin described above. The three fields shown in Figs 5, 6 & 7 were selected to represent contrasting slope angles. These range from an average slope of approximately 12° in the case of field 1, through 6° for field 2, to 3° for field 3. All three fields have been used for arable cultivation during a large proportion of the preceding 25 years. The sampling programme employed involved the collection of samples along downslope transects located so as to afford a near-uniform distribution of sample sites within each field. Additional sampling sites were located in the vicinity of downslope hedge boundaries where depositional areas might be expected. The whole-core samples (42 cm^2) were collected to a depth of approximately 60 cm using a steel tube percussion corer specially developed for this purpose. The basal 2.5 cm of each core was analysed separately for ^{137}Cs content, a zero concentration indicating that the core had penetrated to the bottom of the ^{137}Cs profile.

The data assembled for the fields have been plotted on Figs 5, 6 & 7. In intepreting these data it is important to recognize that the results obtained from the gamma spectrometry analysis involve an analytical precision of the order of ±5%. The total activity recorded for an individual site must therefore differ from the input or control value by more than 5% if it is to afford definitive evidence of erosional loss or depositional gain. In the case of field 1 (Fig. 5), all sample sites, with the exception of those at the base of the slope, provide evidence of erosion. This erosion has clearly been relatively severe, since the ^{137}Cs levels indicate that >50% of the input has been lost from an appreciable proportion of the sites and that the loss is greater than 60% at several sites. Similarly, it is apparent that substantial deposition has taken place at the base of the slope, since samples from these sites evidence ^{137}Cs levels up to 88% greater than the control value. An attempt has been made to generalize the pattern demonstrated by the individual sites by interpolating isopleths through the data points. Such isopleths have been referred to as isocaes by Longmore *et al.* (1983). More sample sites would be required to generate a definitive map, but if it assumed that the magnitude of the erosional loss is approximately proportional to the percentage ^{137}Cs loss, the resultant pattern of erosion indicates that most erosion has occurred on the steepest part of the field and that this has been preferentially concentrated down a spur. Field evidence suggests that this spur area, and also the other two zones evidencing high rates of erosion are characterized by the thinnest soils within the field. The pattern of erosion may thus reflect the increased incidence of surface runoff from these areas. The continuation of the zone of maximum erosion associated with the spur from the top to the bottom of the field suggests that little or no deposition occurs down the slope. This conclusion is substantiated by the absence of areas of enhanced ^{137}Cs concentrations from all areas of the field except the base of the slope.

An approximate ^{137}Cs budget can be derived for the field by estimating the fraction of the overall ^{137}Cs input that has been mobilized by erosion,

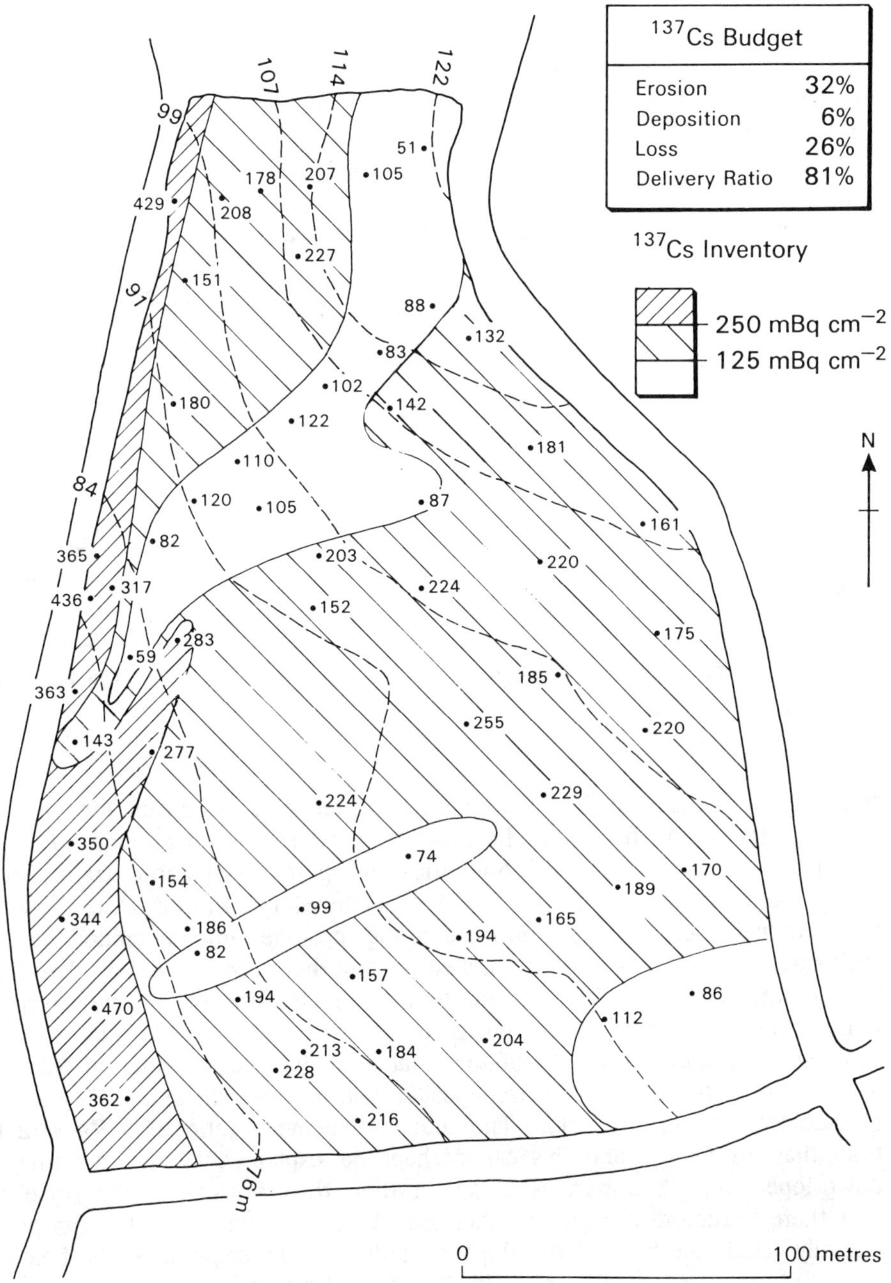

Fig. 5 The distribution of ^{137}Cs within field 1 and the associated ^{137}Cs budget.

and the relative proportions of this fraction that have been redeposited at the base of the slope or transported beyond the field. In this case, approximately 32% of the input of ^{137}Cs has been mobilized, with 7% having been redeposited at the bottom of the slope and the remaining 26% transported beyond the field. Although these proportions relate specifically to the input of ^{137}Cs, it is possible to argue that they are at least indicative of the relative proportions of the eroded soil that have been redeposited or transported beyond the field and therefore of the sediment delivery ratio (cf. Walling *et al.*, 1986a). Following these assumptions, the sediment delivery ratio for field 1 can be estimated at 81%.

Equivalent results for field 2 are presented in Fig. 6. The pattern is

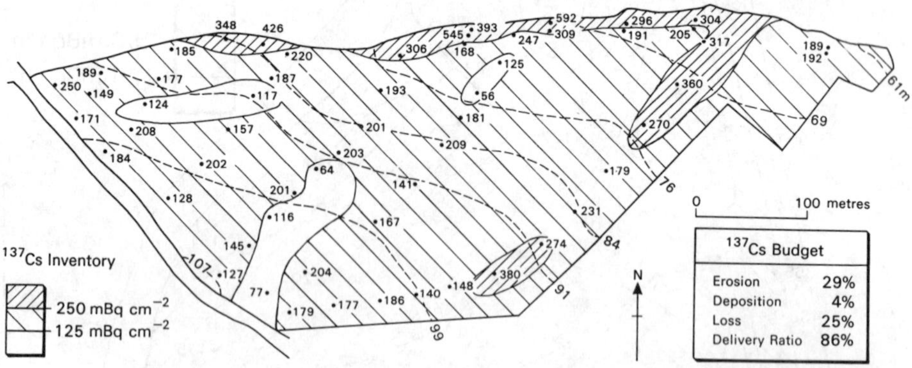

Fig. 6 The distribution of ^{137}Cs within field 2 and the associated ^{137}Cs budget.

similar to that for field 1 in that there is evidence of significant erosion over the majority of the field and the areas of most severe erosion are primarily associated with the contour convexities or spurs. Deposition again occurs above the boundary hedge at the base of the slope. In this case, however, there is evidence of deposition extending upslope into a minor contour concavity which appears to be linked to another small area of deposition towards the top of the field. In addition, the main zone of maximum erosion is not continuous over the full length of the slope. The ^{137}Cs budget for the field again indicates that 29% of the total input has been mobilized, although in this case only 4% has been redeposited within the field and 25% has been transported beyond the field. The sediment delivery ratio of 86% is higher than that for field 1 and this can perhaps be explained by the fact that the downslope hedge boundary is oblique rather than normal to the slope and that there is therefore a greater likelihood that sediment could be transported laterally across the base of the slope towards the lowest point of the field.

The data collected from field 3 and mapped in Fig. 7 evidence a very different pattern to that shown by the previous two examples. Firstly, there are only very small areas of the field where the ^{137}Cs levels are less than 50% of the input or control value and there appears in general to have been less mobilization of the ^{137}Cs input. The areas of maximum ^{137}Cs loss are, however, still associated with areas of contour convexity. Secondly, there are substantial areas of deposition *within* the field as well as along the downslope boundary. Most of the depositional areas within the field are associated with small depressions marked by contour concavities, and it is apparent that a

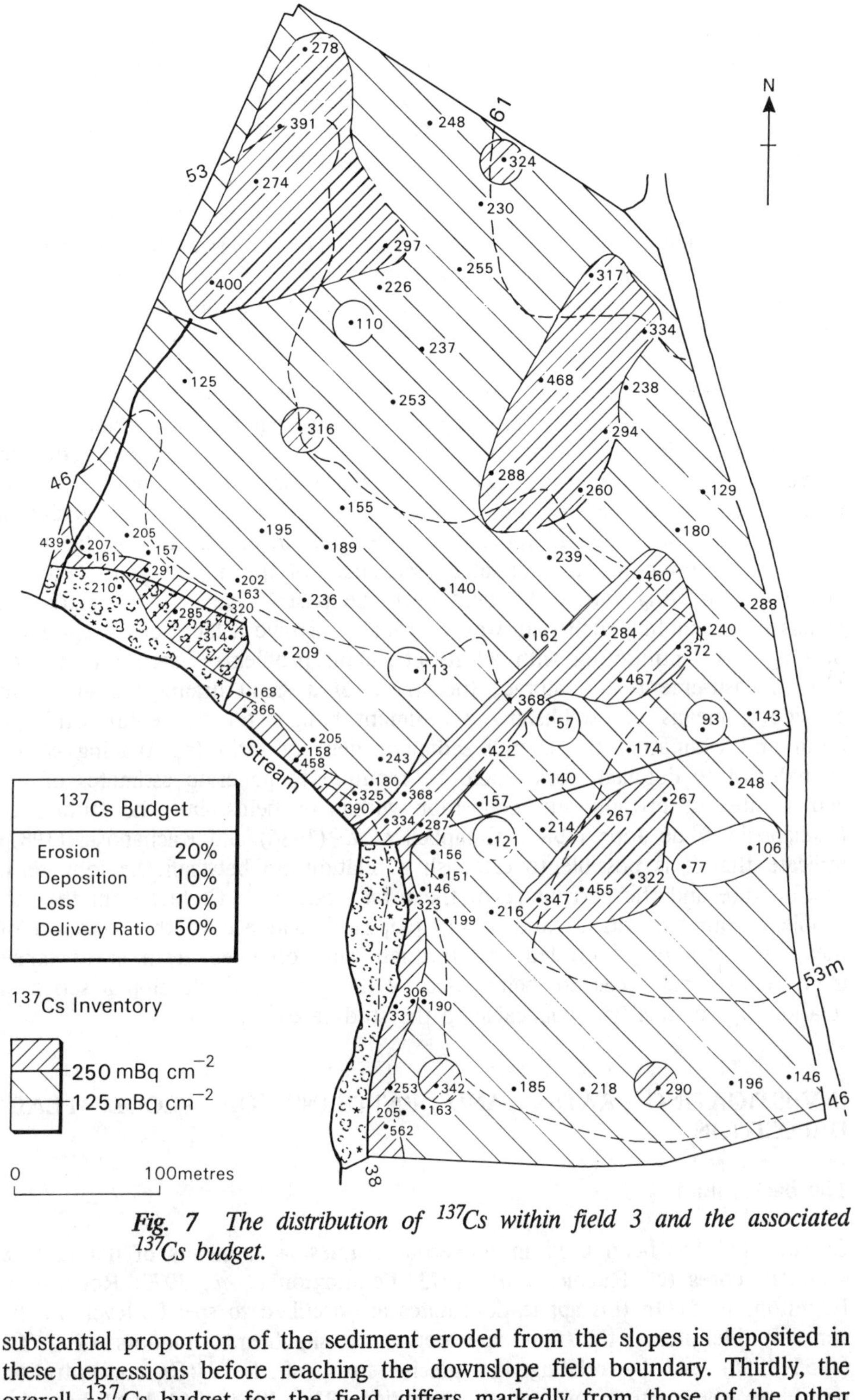

Fig. 7 The distribution of ^{137}Cs within field 3 and the associated ^{137}Cs budget.

substantial proportion of the sediment eroded from the slopes is deposited in these depressions before reaching the downslope field boundary. Thirdly, the overall ^{137}Cs budget for the field differs markedly from those of the other two fields. Only 20% of the total input of ^{137}Cs has been mobilized, and of this 10% has been redeposited within the field and only 10% has been transported beyond the field. The delivery ratio of approximately 50% stands in marked contrast to the values of 81% and 86% associated with the other fields. These contrasts undoubtedly reflect the influence of slope steepness, since the slopes in field 3 are considerably gentler than those in fields 1 and 2. In view of the close similarities between the patterns of erosion and deposition and the sediment delivery ratios associated with fields 1 and 2, it is suggested that similar fields in this locality with slopes of the order of 6° or more are associated with relatively high rates of erosion and efficient sediment delivery systems, whereas fields with slopes of about 3° or less are characterized by much lower erosion rates and reduced sediment delivery ratios.

The results obtained from this study of three cultivated fields demonstrate the potential offered by ^{137}Cs measurements for elucidating the pattern of erosion and deposition within a field and the overall sediment delivery ratio. Although the approach necessarily possess limitations in terms of the representativeness of the sampling sites, the precision of the analytical measurements and the interpolation procedures involved, it has the great advantage of providing an integrated assessment of the impact of the erosion processes operating within the field over the past 25 years. Any alternative means of assembling such data would probably involve a laborious programme of long term monitoring with all its attendant problems. Extension of such ^{137}Cs measurements to embrace the whole of a small drainage basin could provide a means of establishing a sediment budget for the entire drainage basin and estimating the overall sediment delivery ratio (cf. Walling *et al.*, 1986a). More detailed work would be required to generate estimates of the actual rates of erosion and deposition within th fields, but the studies of Kachanoski & de Jong (1984), Campbell *et al.* (1986) and Kachanoski (1987) indicate that it is possible to establish a relationship between the long term erosion rate and the percentage of the total input of ^{137}Cs lost from the soil profile. Both theoretical models and empirical evidence of the variation of the ^{137}Cs content of eroded soil through time obtained from cores taken from depositional areas are being used by the authors to develop a series of relationships of this type applicable to the study area.

INVESTIGATING RATES AND PATTERNS OF FLOOD PLAIN DEPOSITION

The background

Caesium-137 has been used in numerous studies as a means of dating lake sediment cores (cf. Ritchie *et al.*, 1973; Pennington *et al.*, 1973; Robbins & Edginton, 1975). In this application, dates are ascribed to specific levels in the core by identifying the first appearance of significant amounts of ^{137}Cs (1955–1956) and by relating the precise form of the ^{137}Cs concentration profile to the pattern of annual deposition. The 1963–1964 level has, for example, been identified by the existence of a pronounced peak in the concentration values corresponding to the peak fallout at that time (cf. Fig. 1). It must, however, be recognized that a close correspondence between the ^{137}Cs profile and the pattern of annual fallout can only be expected in water bodies where autochthonous sedimentation is dominant, since allochthonous sediment eroded from the watershed will be characterized by ^{137}Cs concentrations reflecting the cumulative pattern of fallout receipt at its source (cf. Walling *et al.,* 1986a). Furthermore, several studies have demonstrated that diffusion and mixing processes and sediment focussing may complicate the ^{137}Cs profiles exhibited by lake sediment cores (cf. Robbins & Edgington, 1975; Davis *et al.,* 1984; Anderson *et al.,* 1987). Detailed investigations of the total ^{137}Cs content (mBq cm^{-2}) of a network of sediment cores from a lake or reservoir could also provide a basis for investigating the spatial pattern of deposition within a water body and the incidence of sediment focussing (e.g. Plato & Jacobson, 1976).

Caesium-137 has also been used to estimate rates of sedimentation in other depositional environments, including salt marshes (Delaune *et al.,* 1978) and river backwater zones (Ritchie & McHenry, 1985). Furthermore, work undertaken by the authors indicates that this radionuclide can provide a basis for documenting rates and patterns of flood plain accretion by overbank deposition of fine sediment. In the case of flood plains, however, it is necessary to take account of both the accumulation and vertical redistribution of ^{137}Cs fallout in the flood plain soils and the addition ^{137}Cs associated with deposited sediment, in interpreting the evidence provided by this radionuclide. Fig. 8 compares two ^{137}Cs profiles (B & C), obtained from permanent pasture located on the flood plain of the River Culm in Devon, UK, with a characteristic profile (A) from permanent pasture in the vicinity of the

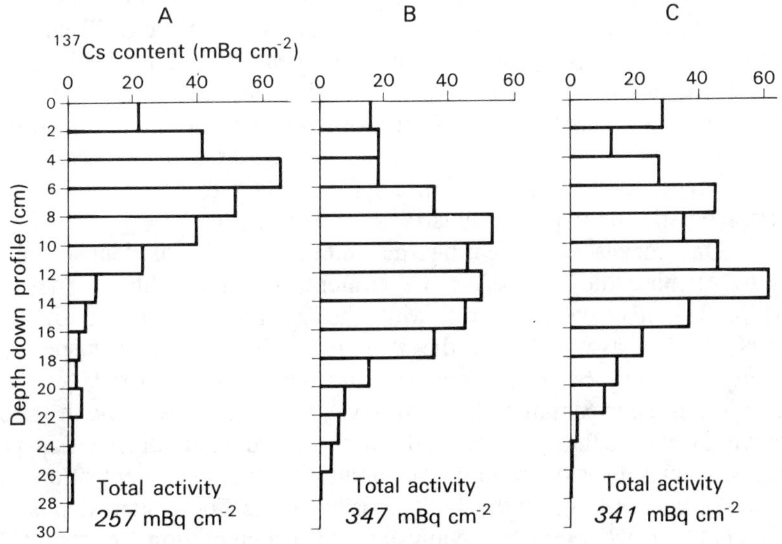

Fig. 8 Caesium-137 profiles representative of an undisturbed input or control site adjacent to the flood plain of the River Culm (A), and of two sites on the flood plain evidencing deposition (B & C).

flood plain, but above the level of flood water inundation. In all cases the profiles represent undisturbed conditions with no cultivation. Profile A may be viewed as a reference or control site. It exhibits the typical exponential depth distribution associated with such sites and the total ^{137}Cs content is consistent with the input values encountered in the neighbouring Jackmoor Brook catchment discussed previously. Profiles B & C from the two flood plain sites are characterized by higher total ^{137}Cs contents and by different depth distributions. Both features are consistent with a situation where sediment deposited during overbank flooding has provided an additional input of ^{137}Cs and has caused an upward "stretching" of the profile. Estimates of deposition rates could be based either on the amount of excess ^{137}Cs or the degree of "stretching" (cf. Walling *et al.,* 1986b).

A case study

Further discussion of the potential for using ^{137}Cs as a means of investigating rates and patterns of flood plain deposition is best undertaken by more detailed reference to work undertaken on the flood plain of the River Culm. Figure 9 depicts a 13 km reach of the lower course of the River Culm, Devon, UK, which has a drainage area of 276 km^2. Along most of this reach, the river flows in a gravel bed channel which is approximately 12 m wide. The banks are up to 1 m high and are largely formed of fine alluvial material. Overbank flooding is relatively frequent during the winter months and substantial inundation of this area of flood plain generally occurs on about seven occasions each year. Depths of inundation vary with the local topography of the flood plain, but in the middle reaches flood water depths are typically about 40 cm for the mean annual flood and 70 cm for a 50 year flood. Caesium-137 profiles have been measured at 8 sites along this reach and these are presented in Fig. 9. In all cases, the total ^{137}Cs content of the flood plain sediments at the profile sites exceeds the input or control value of 260 mBq cm^{-2} and therefore suggests that deposition has occurred. Similarly, all profiles show some evidence of "stretching" when compared with the characteristic input profile illustrated in Fig. 8(A).

One means of estimating the amount or rate of deposition at each site is to examine the "excess" ^{137}Cs contents of the profiles. These range from 81 to 377 mBq cm^{-2}. If, following the arguments advanced by Walling *et al.* (1986b), it is assumed that the average ^{137}Cs content of suspended sediment transported by the river over the past 30 years adjusted for decay to the present, is approximately 12.5 mBq g^{-1} and that this value is also representative of the sediment deposited on the flood plain during this period, these excess values provide tentative estimates of sediment deposition of the order of 6–30 g cm^{-2}. Since the bulk density of the flood plain deposits is typically about 1 g cm^{-3}, these are equivalent to sedimentation depths of 6 to 30 cm and of sedimentation rates of 2 to 10 mm year^{-1} over the past 30 years during which appreciable levels of ^{137}Cs would have been present in suspended sediment.

Comparison of the ^{137}Cs profiles illustrated in Fig. 9 with the input

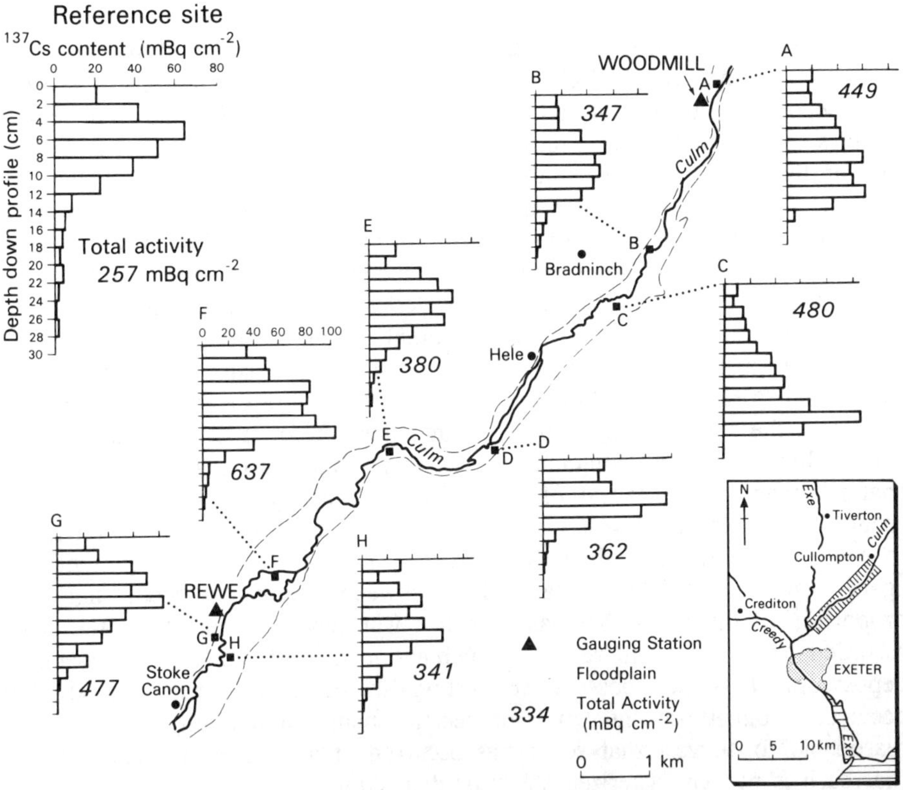

Fig. 9 Caesium-137 profiles associated with eight sites representative of depositional environments along the flood plain of the River Culm.

profile depicted in Fig. 8(A), and also included on Fig. 9, provides a basis for estimating the degree of "stretching" of the profile, which in turn provides an alternative estimate of the depth of sedimentation and the associated sedimentation rate. Matching of the lower portions of the profiles in Fig. 9 with that shown in Fig. 8(A) suggests that the former exhibit "stretching" by of the order of 6 to 20 cm, which is equivalent to sedimentation rates of 2 to 6.7 mm year^{-1}. These values are in reasonably close agreement with those obtained above and the profiles exhibiting high degrees of "stretching" are commonly those with high values of ^{137}Cs "excess". There are some minor discrepancies between the two sets of data, but these are not unexpected in view of the various assumptions involved and the fact that sedimentation may not have been continuous over the period and that the particle size composition and therefore the ^{137}Cs content of the deposited sediment could be expected to vary from site to site. Both approaches to the use of ^{137}Cs measurements to estimate depths and rates of flood plain deposition provide estimates which are consistent with other evidence available for this reach of the River Culm from measurements of the reduction in suspended sediment loads through the reach and from sedimentation traps (cf. Walling *et al.*, 1986b; Lambert & Walling, 1987). Further work is in progress to refine these approaches to estimating depths and rates of flood plain deposition.

As with the use of ^{137}Cs measurements to investigate patterns of erosion and deposition in agricultural fields discussed previously, the number of samples requiring analysis precludes the investigation of a large number of site profiles, in order to study spatial patterns of flood plain deposition. However, single whole-core samples can again be used for this purpose, since the values of total ^{137}Cs content (mBq cm^{-2}) associated with each core may be compared with the input or control value to assess the depth and rate of deposition at that site. Figure 10 provides a map of the study reach onto which the values of total ^{137}Cs content obtained for more than 120 whole-core samples taken from the reach have been superimposed. In this case about 30% of the values are less than the reference value of 260 mBq cm^{-2} and these are thought to represent sites where scour has occurred or where flood plain development has occurred only recently as a result of channel migration. Removal of part of the profile by erosion will result in reduced values of total ^{137}Cs content, and, if deposition has been restricted to only part of the past 30 years, reduced values of total ^{137}Cs content will again occur. No simple relationship between depth of scour and the proportion of the ^{137}Cs that has been lost will, however, exist since the value recorded for a site could reflect a combination of both scour and subsequent deposition. It is not possible to distinguish those sites where scour has occurred from those indicative of recent channel migration, but it seems reasonable to suggest that scour has occurred at a number of sites and that the reach is not characterized solely by deposition.

The values of ^{137}Cs "excess" associated with those sites where the core value exceeds the reference level, and where net deposition has therefore occurred, range up to 773 mBq cm^{-2}. Values of the order of 40–125 mBq cm^{-2} are, however, typical of most of the reach. Based on the assumptions outlined above, these are indicative of depositional depths and rates between approximately 3 and 10 cm and 1 and 3 mm year^{-1} respectively. Looking at the overall distribution of values within the reach there would also appear to be a tendency for the value of ^{137}Cs excess, and therefore the depths and rates of deposition, to increase downstream in response to the increasing width and decreasing gradient of the flood plain. In the downstream portion of the reach in the vicinity of the villages of Rewe and Stoke Canon, several values of ^{137}Cs excess greater than 400 mBq cm^{-2} are evident. These are indicative of depositional depths and rates in excess of about 30 cm and 10 mm year^{-1} respectively.

The whole-core approach can also be used to investigate local patterns of sedimentation within smaller areas of flood plain. Two examples of such studies undertaken within the study reach of the River Culm are provided in Fig. 11. In this case the objective was to investigate the influence of the microtopography of the flood plain on depths and rates of sedimentation. At the first location (Fig. 11(A)), attention focussed on the influence of a small, essentially closed, depression about 50 m from the river channel. The values of total ^{137}Cs content for the individual cores plotted on the map indicate that relatively high rates of deposition occurred immediately adjacent to the channel and also in the bottom of the depression. Depths and rates of

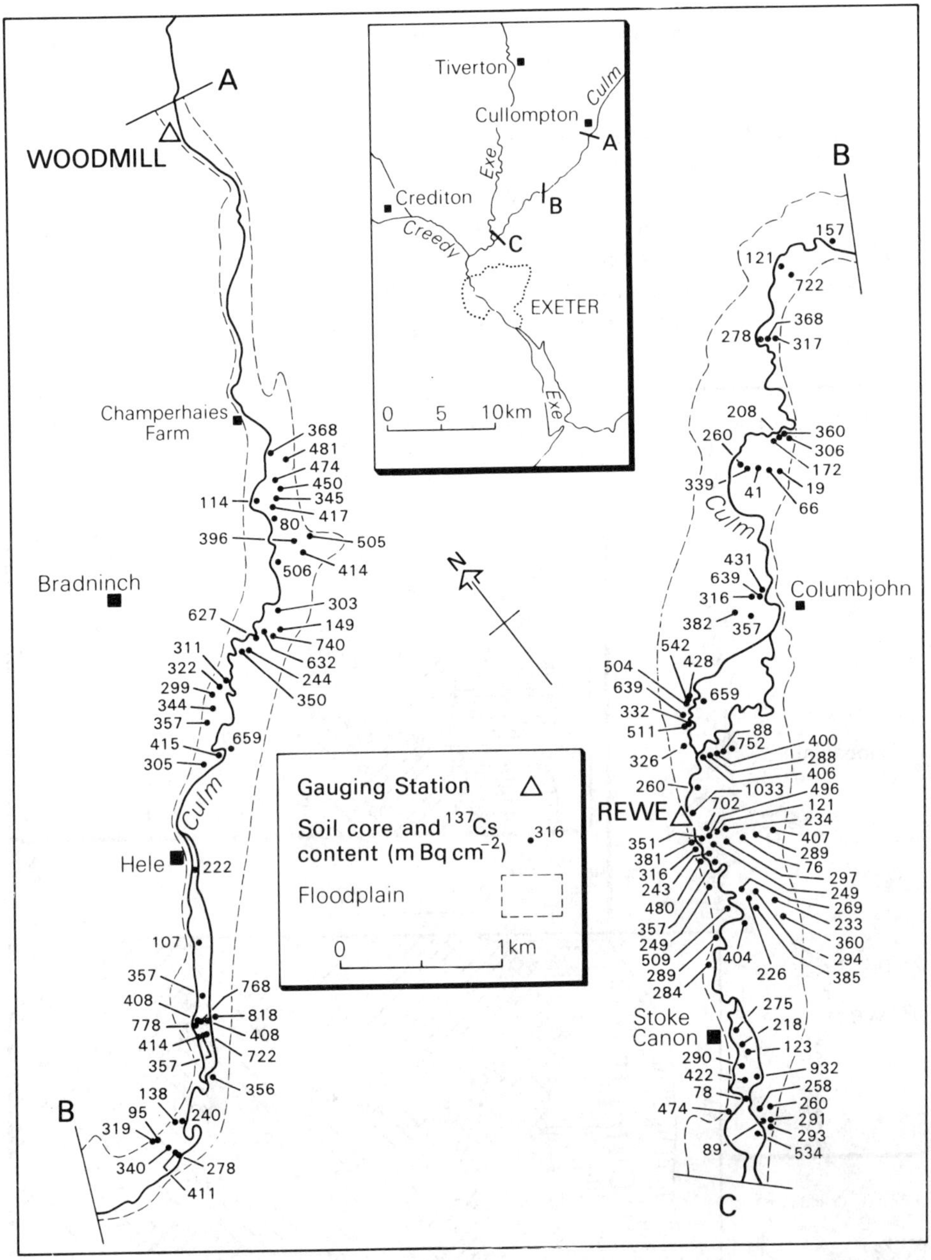

Fig. 10 Values of ^{137}*Cs content associated with whole-core samples collected from a variety of locations along the flood plain of the River Culm.*

deposition in the former location can be estimated to be of the order of 10–40 cm or 3–13 mm year^{-1}. In the centre of the depression the corresponding values are 36 cm and 12 mm year^{-1}. The existence of several points where ^{137}Cs totals are less than the reference value of 260 mBq cm^{-2}

points to the local occurrence of scour which would appear in this case to be associated with the small chutes entering the depression to the north and west.

At the second study site (Fig. 11(B)), interest focussed on the influence of a linear depression trending essentially parallel to the channel. In this case there is no evidence of relatively high rates of deposition adjacent to the channel, and most of the area within and surrounding the depression evidences scour or only relatively low values of ^{137}Cs "excess". The depression apparently acts as a secondary conveyance channel during overbank flows and the relatively high flow velocities occurring at such times cause scour, or at

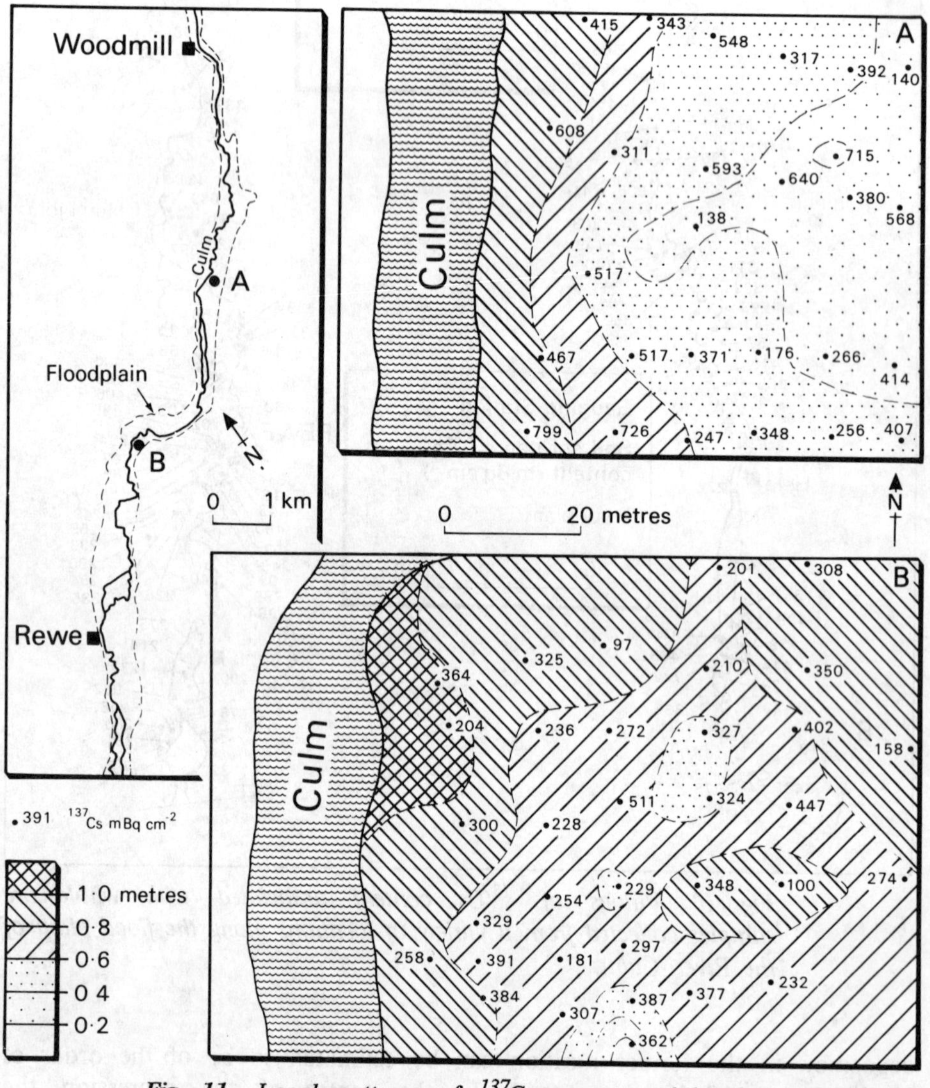

Fig. 11 Local patterns of ^{137}Cs content exhibited by whole-core samples collected from two small areas of the flood plain of the River Culm adjacent to the river channel.

least severely limit rates of deposition. There is some evidence of increased deposition in the more pronounced part of the depression to the south and this may be ascribed to the isolation of this area as a small closed depression during the latter stages of flood water drainage from the flood plain. The considerable spatial variability in deposition rates evident within these two small study areas emphasizes the complexity of flood plain aggradation and suggests that scope exists to use even more intensive sampling networks to unravel the patterns involved and their controls.

THE PROSPECT

The three examples of the use of ^{137}Cs measurements in the study of fluvial erosion, transport and deposition described in this paper undoubtedly represent only a few of the wide range of potential applications. More work is required to investigate other applications and to exploit this potential fully. With its half-life of 30.1 years, ^{137}Cs originating from the bomb tests undertaken in the late 1950s and the 1960s will provide the geomorphologist with a valuable tracer for several decades to come and the opportunities that it offers should be grasped. In those areas where significant amounts of fallout occurred as a result of the Chernobyl accident, interpretation may prove more difficult, but additional opportunities could exist to trace the movement of this essentially instantaneous input of ^{137}Cs through the fluvial system (cf. Walling and Bradley, 1988).

Acknowledgement The case studies described in this paper were undertaken by the authors as part of a wider study of sediment delivery processes financed by the UK Natural Environment Research Council (NERC). This support and the generous assistance of many local landowners in permitting access to field sampling sites is gratefully acknowledged. The authors have also benefitted from a number of discussions on the application of ^{137}Cs measurements with Mr B. L. Campbell of the Australian Nuclear Science and Technology Organization and Professor R. J. Loughran of the University of Newcastle, New South Wales, Australia.

REFERENCES

Anderson, R. F., Schiff, S. L., & Hesslein, R. H. (1897) Determining sediment accumulation and mixing rates using ^{210}Pb, ^{137}Cs, and other tracers: problems due to postdepositional mobility or coring artifacts. *Can. J Fish. Aquat. Sci.* **44**, 231-250.

Beasley, D. B., Huggins, L. F. & Monke, E. H. (1982) Modelling sediment yields from agricultural watersheds. *J. Soil Wat. Conserv.* **37**, 114-117.

Cambray, R. S., Fisher, E. M. R., Playford, K., Eakins, J. D. & Peirson, D. H. (1980) Radio-active fallout in air and rain: results to end of 1979. UK Atomic Energy Authority Rep. AERE-R- 9672.

Cambray, R. S., Playford, K. & Lewis, G. N. J. (1982) Radioactive fallout in air and rain: results to end of 1981. UK Atomic Energy Authority Rep. AERE-R-10845.

Cambray, R. S., Cawse, P. A., Garland, J. A., Gibson, J. A. B., Johnson, P., Lewis, G. N. J., Newton, D., Salmon, L. & Wade, B. O. (1987) Observations on radioactivity from the

Chernobyl accident. *Nuclear Energy* **26**, 77-101.
Campbell, B. L., Loughran, R. J. & Elliott, G. L. (1982) Caesium-137 as an indicator of geomorphic processes in a drainage basin system. *J. Aust. Geogr. Studies* **20**, 49-64.
Campbell, B. L. (1983) Application of environmental caesium-137 for the determination of sedimentation rates in reservoirs and lakes and related catchment studies in developing countries. In: *Radioisotopes in Sediment Studies*. IAEA-TECDOC-298, 7-29.
Campbell, B. L., Loughran, R. J., Elliott, G. L. & Shelly, D. J. (1986) Mapping drainage basin sediment sources using caesium-137. In: *Drainage Basin Sediment Delivery* (Proc. Albuquerque Symp. Aug. 1986), 437-446 IAHS Publ. no. 159.
Davis, M. B., Moeller, R. E. & Ford, J. (1984) Sediment focusing and pollen influx. In: *Lake Sediments and Environmental History* (ed. E. Y. Haworth & J. W. Lund), Leicester Univ., UK, 261-293.
De Jong, E., Begg, C. B. M. & Kachanoski, R. G. (1983) Estimates of soil erosion and deposition for some Saskatchewan soils. *Can. J. Soil Sci.* **63**, 607-617.
Delaune, R. D., Patrick, W. H. & Buresh, R. J. (1978) Sedimentation rates determined by ^{137}Cs dating in a rapidly accreting salt marsh. *Nature* **275**, 532-533.
Dorr, H. & Munnich, K. O. (1987) Spatial distribution of soil- ^{137}Cs and ^{134}Cs in West Germany after Chernobyl. *Naturwissenschaften* **74**, 249-251.
Finlayson, B. L. (1978) Suspended solid transport in a small experimental catchment. *Z. Geomorphol.* **NF22**, 192-210.
Frissel, M. J. & Pennders, R. (1983) Models for the accumulation and migration of ^{90}Sr, ^{137}Cs, ^{239,240}Pu and ^{241}Am in the upper layer of soils. In: *Ecological Aspects of Radionuclide Release* (ed. P. J. Coughtrey), 63-72, Special Publ. Brit. Ecol. Soc. No. 3.
Gale, H. J., Humphreys, D. L. O. & Fisher, D. M. R. (1963) The weathering of caesium-137 in soil. UK Atomic Energy Authority Rep. AERE-R-4241.
Glymph, L. M. (1975) Evolving emphases in sediment yield predictions. In: *Present and Prospective Technology of Predicting Sediment Yields and Sources*. USDA ARS-S-40, 1-4.
Kachanoski, R. G. (1987) Comparison of measured soil 137-cesium losses and erosion rates. *Can. J. Soil Sci.* **67**, 199-203.
Kachanoski, R. G. & De Jong, E. (1984) Predicting the temporal relationship between soil cesium-137 and erosion rate. *J. Environ. Qual.* **13**, 301-304.
Lambert, C. P. & Walling, D. E. (1987) Floodplain sedimentation. A preliminary investigation of contemporary deposition within the lower reaches of the River Culm, Devon, UK. *Geograf. Ann.* **69A**, 47-59.
Longmore, M. E., O'Leary, B. M., Rose, C. W. & Chandica, A. L. (1983) Mapping soil erosion and accumulation with the fallout isotope caesium-137. *Aust J. Soil Res.* **21**, 373-385.
Loughran, R. J., Campbell, B. L. & Elliott, G. L. (1982) The identification and quantification of sediment sources using ^{137}Cs. In: *Recent developments in the Explanation and Prediction of Erosion and Sediment Yield* (Proc. Exeter Symp. July 1982), 361-369. IAHS Publ. no. 137.
McHenry, J. R. & Ritchie, J. C. (1977) Estimating field erosion losses from fallout caesium-137 measurements. In: *Erosion and Solid Matter Transport in Inland Waters*. (Proc. Paris Symp. July 1977), 26-33, IAHS Publ. no. 122.
McHenry, J. R., Ritchie, J. C. & Bubenzer, G. D. (1978) Redistribution of caesium-137 due to erosional processes in a Wisconsin watershed. In: *Environmental Chemistry and Cycling Processes* (Proc. Augusta Conf., May 1976), 495-503, ERDA Symp. Series.
Oldfield, F., Rummery, T. A., Thompson, R. & Walling, D. E. (1970) Identification of suspended sediment sources by means of magnetic measurements. *Wat. Resour. Res.* **15**, 211-218.
Peart, M. R. & Walling, D. E. (1986) Fingerprinting sediment sources: The example of a drainage basin in Devon, UK. In: *Drainage Basin Sediment Delivery* (Proc. Albuquerque Symp. Aug. 1986), 41-55, IAHS Publ. no. 159.
Pennington, W., Cambray, R. S. & Fisher, E. M. (1973) Observations on lake sediment using fallout ^{137}Cs as a tracer. *Nature* **242**, 324-326.
Plato, P. & Jacobson, A. P. (1976) Cesium-137 in Lake Michigan sediments: areal distribution and correlation with other man-made materials. *Environ. Pollut.* **10**, 19-33.
Ritchie, J. C. (1987) Literature relevant to the use of radioactive fallout cesium-137 to measure soil erosion and sediment deposition. USDA ARS Hydrology Laboratory Tech. Rep. HL-9.
Ritchie, J. C., McHenry, J. R. & Gill, A. C. (1973) Dating recent reservoir sediments. *Limnol. Oceanogr.* **18**, 254-263.
Ritchie, J. C. & McHenry, J. R. (1985) A comparison of three methods of measuring recent rates of sediment accumulation. *Wat. Resour. Res.* **21**, 91-103.
Robbins, J. A. & Edgington, D. N. (1975) Determination of recent sedimentation rates in Lake Michigan using Pb-210 and Cs-137. *Geochim. Cosmochim. Acta* **39**, 285-304.
Rogowski, A. S. & Tamura, T. (1970a) Environmental mobility of cesium-137. *Radiation Botany* **10**, 35-45.
Rogowski, A. S. & Tamura, T. (1970b) Erosional behaviour of cesium-137. *Health Phys.* **18**,

467-477.
Swanson, F. J., Janda, R. J., Dunne, T. & Swanston, D. N. (1982) Sediment budgets and routing in forested drainage basins. USDA Forest Service General Tech. Rep. PNW-141.
Tamura, T. (1964) Selective sorption reaction of caesium with mineral soils. *Nuclear Safety* **5**, 262-268.
US Health and Safety Laboratory (1977) Environmental Quarterly, July 1, ERDA Rep. HASL-321
Wall, G. J. & Wilding, L. P. (1976) Mineralogy and related parameters of fluvial suspended sediments in Northwestern Ohio. *J. Environ. Qual.* **5**, 168-173, 35. 1.
Walling, D. E. & Bradley, S. B. (1988) Transport and redistribution of Chernobyl fallout radionuclides by fluvial processes: some preliminary evidence. *Environ. Geochem. Health* **10**, 35-39.
Walling, D. E., Bradley, S. B. & Wilkinson, C. J. (1986a) A caesium-137 budget approach to the investigation of sediment delivery from a small agricultral drainage basin in Devon, UK. In: *Drainage Basin Sediment Delivery,* (Proc. Albuquerque Symp., Aug. 1986), 423-435. IAHS Publ. no. 159.
Walling, D. E., Bradley. S. B. & Lambert, C. P. (1986b) Conveyance losses of suspended sediment within a floodplain system. In: *Drainage Basin Sediment Delivery* (Proc. Albuquerque Symp., Aug. 1986), 119-131, IAHS Publ. no. 159.
Walling, D. E. & Kane, P. (1982) Temporal variations of suspended sediment properties. In: *Recent Developments in the Explanation and Prediction of Erosion and Sediment Yield.* (Proc. Exeter Symp. July 1982) 409-419, IAHS Publ. no. 137.
Wolman, M. G. (1977) Changing needs and opportunities in the sediment field. *Wat. Resour. Res.* **13**, 50-54.
Wood, P. A. (1978) Fine-sediment mineralogy source rocks and suspended sediment, Rother Catchment, West Sussex. *Earth Surf. Processes* **3**, 255-263.

Sediment and the Environment

Edited by R. F. Hadley & E. D. Ongley

price $40 (US)
(including postage by surface mail)
218 + viii pages
IAHS Publication no. 184 *(published May 1989)*

ISBN 0-947571-12-4

This publication contains papers selected for presentation at the ***International Symposium on Sediment and the Environment*** held during the Third Scientific Assembly of IAHS at Baltimore, Maryland, in May 1989. The papers dealt with sediment problems and contaminant transport associated with both natural and disturbed environments. The four main topics were: (1) sediment-asssociated transport of contaminants in nonpoint pollution; (2) erosion control problems associated with mining, construction and waste disposal activities; (3) time lag in sediment movement through drainage networks; and (4) the modelling of runoff and sedimentation.

Among the papers on topic (1), one that attracted particular interest was on the fallout of radionuclides from the Chernobyl reactor accident in 1986. Measurements of the caesium^{137} content of suspended sediment in the River Severn in the UK were used to assess the role of fluvial transport in the distribution of radionuclides. A further complication to the problem of assessing pollution associated with sediment transport was discussed in a paper on samplers. Sampling and subsequent physical and chemical analyses of suspended sediment from various locations in the USA indicated substantial differences in sediment concentrations and some trace elements depending on the type of sampler used — depth-integrated, point or pumping sampler.

Papers dealing with topic (2) emphasized cost-impacts of soil erosion on storage reservoirs and the problems caused by floods with hyper-concentrations of sediment. Also, the need for design and management of reclaimed lands in order to re-establish some degree of equilibrium when geomorphic systems are disturbed was considered.

The processes of long-term and short-term storage of sediment in hydrological systems, and the use of lead and zinc trace metals associated with mining are shown to be useful in estimating sedimentation rates.

Orders This IAHS Publication may be ordered from the following addresses:

**Office of the Treasurer
IAHS, 2000 Florida
Avenue NW, Washington
DC 20009, USA**
[telephone: 202 462 6903]

**Bureau des Publications de
l'UGGI, 140 Rue de Grenelle
75700 Paris, France**
*[téléphone: 45 50 34 95 ext. 816;
telex: 204989 igngnl f (Attn: UGGI)]*

**IAHS Press, Institute of
Hydrology, Wallingford
Oxfordshire OX10 8BB, UK**
*[telephone: (0)491 38800;
telex: 849365 hydrol g]*

NTERNATI(HYDROLOGICAL SCIENCES

IAI Geodesy and Geophysics

eedings and R...

over 60 years the Association has shed proceedings of symposia and shops, and reports from working ps. Such publications comprise the established "red book" series of eedings and Reports, and these may dered from:

of the Treasurer IAHS, 2000 Florida Avenue Vashington, DC 20009, USA

Press, Institute of Hydrology, Wallingford, dshire OX10 8BB, UK

ographs and Reports

989 the first of a new series of ographs and Reports was published. series is being published in eration with the International ute for Hydraulic and Environmental neering, Delft, The Netherlands. s in this series will have an A4 at, will not exceed 100 pages, and may rdered only from the Wallingford ess given above.

logue of Publications

atalogue of all IAHS publications includes information about the nal may be obtained from either Washington or the Wallingford ess given above.

s of the Association

ie objects of the Association are:

) To promote the study of Hydrology as an of the earth sciences and of water resources; dy the hydrological cycle on the earth and the s of the continents; the surface and ground s, snow and ice, including their physical, cal and biological processes, their relation to e and to other physical and geographical s as well as the interrelations between them; dy erosion and sedimentation and their n to the hydrological cycle; to examine the logical aspects of the use and management of resources and their change under the ce of man's activities; to provide a firm ific basis for the optimal utilization of water ces systems, including the transfer of edge on planning, engineering, management onomic aspects of applied hydrology.

To provide for discussion, comparison, and ation of research results.

To initiate, facilitate, and coordinate research nd investigation of, those hydrological prob- hich require international cooperation.

The Association is a constituent body of the International Union of Geodesy and Geophysics. The Association is subject to those articles of the statutes and bye-laws of the Union that apply to associations and also to these statutes.

3. Any country adhering to the Union also adheres to the Association, and is entitled to send delegates and otherwise to participate in its work. All scientific meetings of the Association or of its components are open to such delegates.

4. The Association performs its scientific and international activities in the framework of the International Council of Scientific Unions and IUGG in cooperation with the United Nations and its specialized agencies and through direct contacts with other international organizations.

5. The Association shall comprise: the General Assembly, the Bureau of the Association, the Commissions, and the Committees. The Association maintains contact with the several adhering countries through their National Committees (subcommittees or sections of the IUGG National Committees or national representatives).

Commissions: Units having defined responsibilities in specific hydrological subjects. Divisions are corresponding units of the Commissions.

Committees: Units of the Association, ad hoc or standing, for scientific and administrative purposes, and are areally differentiated units such as: National Committees and regional groups of National Committees.

Panels or Working Groups: Ad hoc units to report on specific problems, either scientific or administrative.

International Hydrology Prize

Since 1981 an International Hydrology Prize has been awarded annually on an individual basis in recognition of an outstanding contribution to the science. Nominations for the prize are made by National Committees and forwarded to the Secretary General for consideration by the Nominations Committee. Details of the criteria considered for the award are available from the Secretary General.

Tison Award

The Tison Fund was established in 1982 to provide an annual prize of $750 (US). The Tison Award will be granted for an outstanding paper published by IAHS in a period of two years previous to the deadline for nominations. **Nominations should be received by the Secretary General not later than 31 December each year.** Candidates must be **under 41 year of age** at the time their paper was published. The full list of rules may be obtained from the Secretary General.

INTERNATIONAL ASSOCIATION OF HYDROLOGICAL SCIENCES